WITHDRAWN FROM LIBRARY

Neuroimaging and Neurophysiology in Psychiatry

D1423657

WITHDRAWN FROM LIBRARY

BRITISH MEDICAL ASSOCIATION

0827109

Neuroimaging and Neurophysiology in Psychiatry

David E.J. Linden
Division of Psychological Medicine and Clinical Neurosciences, School of Medicine, Cardiff University, Cardiff, UK
Cardiff University Brain Research Imaging Centre, School of Psychology, Cardiff University, Cardiff, UK

OXFORD
UNIVERSITY PRESS

Great Clarendon Street, Oxford, OX2 6DP,
United Kingdom

Oxford University Press is a department of the University of Oxford.
It furthers the University's objective of excellence in research, scholarship,
and education by publishing worldwide. Oxford is a registered trade mark of
Oxford University Press in the UK and in certain other countries

© Oxford University Press 2016

The moral rights of the author have been asserted

First Edition published in 2016
Impression: 1

All rights reserved. No part of this publication may be reproduced, stored in
a retrieval system, or transmitted, in any form or by any means, without the
prior permission in writing of Oxford University Press, or as expressly permitted
by law, by licence or under terms agreed with the appropriate reprographics
rights organization. Enquiries concerning reproduction outside the scope of the
above should be sent to the Rights Department, Oxford University Press, at the
address above

You must not circulate this work in any other form
and you must impose this same condition on any acquirer

Published in the United States of America by Oxford University Press
198 Madison Avenue, New York, NY 10016, United States of America

British Library Cataloguing in Publication Data
Data available

Library of Congress Control Number: 2015949835

ISBN 978–0–19–873960–9

Printed and bound by
CPI Group (UK) Ltd, Croydon, CR0 4YY

Oxford University Press makes no representation, express or implied, that the
drug dosages in this book are correct. Readers must therefore always check
the product information and clinical procedures with the most up-to-date
published product information and data sheets provided by the manufacturers
and the most recent codes of conduct and safety regulations. The authors and
the publishers do not accept responsibility or legal liability for any errors in the
text or for the misuse or misapplication of material in this work. Except where
otherwise stated, drug dosages and recommendations are for the non-pregnant
adult who is not breast-feeding

Links to third party websites are provided by Oxford in good faith and
for information only. Oxford disclaims any responsibility for the materials
contained in any third party website referenced in this work.

Preface

Neuroimaging and neurophysiology are the main non-invasive techniques for the investigation of brain structure and function, and as such play a central role in psychiatric research. They also have important clinical indications in psychiatric practice. This book provides an introduction to the methods of neuroimaging and neurophysiology, describes their clinical indications, and reviews their contribution to the understanding of disease mechanisms and the development of new therapies.

I am grateful to my editor at Oxford University Press, Peter Stevenson, for the original suggestion to write a book on this topic and for his guidance through the writing process. I have received invaluable help in the form of critical comments and illustrations from several experts based at Cardiff University School of Medicine and Cardiff University Brain Research Imaging Centre (CUBRIC). Mark Drakesmith, John Evans, Derek Jones, Thomas Lancaster, Chris Marshall, and Kevin Murphy commented on Chapter 2 and provided original illustrations of imaging methods and results; Krish Singh advised me on Chapter 3; Judith Harrison commented on the chapters on clinical indications of MRI and EEG; Neil Robertson provided advice on the neuropsychiatry of multiple sclerosis; George Kirov provided expertise on EEG monitoring of electroconvulsive treatment; Michael Kerr suggested important background literature on non-epileptic attack disorders; and Daniela Pilz advised me on the section on malformations of cortical development in Chapter 6. All remaining errors and inaccuracies remain my sole responsibility.

I hope that this book will serve as a useful guide to the methods and applications of neuroimaging and neurophysiology for trainees and practitioners in psychiatry and other mental health professions as well as for advanced students and researchers in clinical neuroimaging.

David E.J. Linden

Contents

Abbreviations

5-HT	5-hydroxytryptamine (serotonin)
ACC	anterior cingulate cortex
AD	Alzheimer's disease/axial diffusivity
ADC	apparent diffusion coefficient
ADHD	attention deficit hyperactivity disorder
ANOVA	analysis of variance
APP	amyloid precursor protein
ASD	autism spectrum disorder
ASL	arterial spin labelling
BOLD	blood oxygenation-level dependent
CAT	computer-assisted tomography
CBF	cerebral blood flow
CBT	cognitive behavioural therapy
CBV	cerebral blood volume
CCK	cholecystokinin
CJD	Creutzfeldt Jakob Disease
CNV	copy number variants
CSF	cerebrospinal fluid
CT	computed tomography
DASB	(2-diethylaminomethylphenylsulfanyl)-benzonitrile
DAT	dopamine transporter
DBS	deep brain stimulation
DLB	dementia with Lewy bodies
DNET	dysembryoplastic neuroepithelial tumor
DRI	dopamine reuptake inhibitor
DSC	dynamic susceptibility contrast
DSM	*Diagnostic and Statistical Manual of Mental Disorders*
DTI	diffusion tensor imaging
DWI	diffusion-weighted imaging
EBO	early brain overgrowth
ECT	electroconvulsive treatment
EEG	electroencephalography
EMG	electromyography
EOG	electrooculography
EP	evoked potentials
EPI	echo-planar imaging
ERP	event-related potential
EPs	evoked potentials
FA	fractional anisotropy
FDG	fluoro-deoxyglucose
FDR	false discovery rate
FFA	fusiform face area
fMRI	functional magnetic resonance imaging
FTD	frontotemporal dementia
FXS	fragile-X syndrome
GABA	gamma-aminobutyric acid
GWAS	genome-wide association studies
HARDI	high angular resolution diffusion-weighted imaging

HF	high frequency
HMPAO	hexamethylpropyleneamine oxime
HPA	hypothalamic pituitary adrenal
ICA	independent components analysis
ICD	*The ICD-10 Classification of Mental and Behavioural Disorders*
LF	low frequency
LIS	classical lissencephaly
LM	light microscopy
LSD	lysergic acid diethylamide
MAO	monoamine oxidase
MCD	malformations of cortical development
MCI	mild cognitive impairment
MD	mean diffusivity
MDD	major depressive disorder
MEG	magnetoencephalography
MFB	medial forebrain bundle
MMN	mismatch negativity
MRI	magnetic resonance imaging
MR	magnetic resonance
MRS	magnetic resonance spectroscopy
MVPA	multi-voxel pattern analysis
NAA	N-acetylaspartate
NET	noradrenaline transporter
NIRS	near infrared spectroscopy
NMR	nuclear magnetic resonance
NREM	non-rapid eye movement
OCD	obsessive compulsive disorder
OFC	orbitofrontal cortex
PCA	principal components analysis
PCP	phencyclidine
PD	Parkinson's disease / proton density
PET	positron emission tomography
PH	periventricular heterotopias
PMG	Polymicrogyria
PNES	psychogenic non-epileptic seizures
PPA	parahippocampal place area
PWI	perfusion-weighted imaging
RBD	REM sleep behaviour disorder
rCBF	regional cerebral blood flow
RCT	randomized controlled trial
rCMRGlc	regional cerebral metabolic rate of glucose consumption
RD	radial diffusivity
RDoC	research domain criteria
REM	rapid eye movement
RFE	recursive feature elimination
RF	radiofrequency
RMT	resting motor threshold
ROI	region of interest
rTMS	repetitive transcranial magnetic stimulation
SCR	skin conductance response
SERT	serotonin transporter (5-HTT)
SNPs	single nucleotide polymorphisms

SNRI	selective noradrenaline reuptake inhibitor
SPECT	single photon emission computed tomography
SQUID	superconducting quantum interference device
SSRI	selective serotonin reuptake inhibitor
TBSS	tract-based spatial statistics
tDCS	transcranial direct current stimulation
tACS	transcranial alternating current stimulation
TDRL	temporal difference reward learning
TE	echo time
TMS	transcranial magnetic stimulation
TR	repetition time
VCFS	velocardiofacial syndrome
VD	vascular dementia
VMPFC	ventromedial prefrontal cortex
WBA	whole brain analysis

Chapter 1

Clinical and research uses of neuroimaging and neurophysiology in psychiatry

Key points

- Most psychiatric diagnoses are based on patients' reported symptoms and observed behaviour
- Neuroimaging and neurophysiology are sometimes clinically indicated to rule out recognisable organic pathologies
- They can also sometimes clarify differential diagnoses, for example, between Alzheimer's and vascular dementia
- Neuroimaging and neurophysiology are also central techniques in psychiatric research
- They can help the quest for biomarkers of mental disorders and chart disease progression
- They can also identify neural correlates of psychiatric symptoms and new treatment targets
- Researchers need to be aware of theoretical pitfalls and practical difficulties

1.1 Biological techniques in psychiatry

Psychiatry rarely has biological techniques available to support diagnosis and treatment decisions. Psychiatrists generally have to rely on the patient's appearance and behaviour, the carefully elicited history, the reported symptoms, examination of the patient's mental state, and the collateral information from family, friends, colleagues, or carers. They use their clinical acumen and experience and training in diagnostic interviewing techniques to arrive at a diagnosis, suggest a treatment, and formulate a prognosis. Tests to confirm a diagnosis of the type commonly used in other fields of medicine—biochemical, immunological, physiological, or imaging—are not available or helpful in the vast majority of psychiatric examinations. During the period of transition from the fourth to fifth edition of the American Diagnostic and Statistical Manual of Mental Disorders (DSM-IV to DSM-5, respectively), hopes were running high that the new diagnostic system would incorporate biological criteria and put psychiatric nosology on a more objective and solid footing. This has not happened, simply because the biology of most psychiatric disorders is still largely unknown. Diagnoses have remained descriptive, and the foundations of most treatments are empirical rather than being grounded in biological disease models.

At the same time, it would be highly desirable to unravel the biological mechanisms of alterations of the mind in order to obtain more reliable diagnoses, causal treatment targets and biomarkers for prognosis, and treatment stratification and monitoring. This urgent need for biological models explains the central status of neurophysiology and neuroimaging in psychiatric research—after all, these are the only techniques that allow for the assessment of the structure and function of the living human brain (Table 1.1). Although their clinical use is mainly confined to the exclusion of other 'neurological' causes of mental illness, the techniques of non-invasive neurophysiology and neuroimaging have become central to most research programmes in biological psychiatry. Trainees and practitioners of psychiatry and related disciplines, for example, clinical psychology or mental health nursing, need to know not only about the standard clinical uses of electroencephalography (EEG), computed tomography (CT), and magnetic resonance imaging (MRI), but also the more specialist clinical applications of radioligand techniques and the wide range of current research activities in psychiatric neurophysiology and neuroimaging.

This book aims to provide a grounding in the physical and physiological bases of these methods (Chapters 2 and 3) as well as their clinical (Chapters 4 and 5) and research applications (Chapters 6 to 10), and the emerging fields of forensic imaging and imaging-based therapies.

Table 1.1 Diagram summarizing biological research techniques in psychiatry (those relevant to this book are in italics)

Source of material	Key techniques	Rationale
Blood, Saliva	Genotyping	Discovery of genetic associations
Blood, Saliva, Cerebrospinal fluid (CSF)	Biochemical analysis	Identification of biomarkers (e.g. hormonal, immunological, proteome, lipidome)
Skin, hair follicles	Reprogramming induced pluripotent stem cells	Cellular models of disease processes, models for drug testing
Brain (non-invasive)	*Neuroimaging, neurophysiology*	*Identification of biomarkers and treatment targets, neurobiological models of disease processes*
Brain (post-mortem)	Histology, gene expression analysis	Identification of biomarkers, neurobiological models of disease process
Animal models	Behaviour, biochemistry, genetics, histology, *neuroimaging, neurophysiology* (invasive or *non-invasive*)	*Neurobiological models of disease process, models for drug testing*

1.2 **Borders between psychiatry and neurology**

The clinical applications of neurophysiology and neuroimaging in psychiatry can be broadly divided into two scenarios. In the first scenario, imaging techniques are used to rule out a primary non-psychiatric diagnosis (for example, whether a clinical presentation of psychosis is caused by a brain tumour). In the second, they are used to distinguish between two pathologies that would both fit the clinical picture (for example, between Alzheimer's disease and vascular dementia). In the first case, the outcome of the diagnostic imaging decides whether the patient comes under the primary care of the neurologists and neurosurgeons (brain tumour) or under that of the mental health services (primary psychosis). Here the demarcation line is fairly clear because the pathology detected by imaging is potentially treatable (by surgery, radiotherapy, and/or chemotherapy) and there is a good chance that the psychiatric symptoms will disappear after successful neurological and neurosurgical treatment. In the second case, the distinction between neurological and psychiatric disorders is less clear cut. Although both Alzheimer's and vascular dementia have clearly identifiable biological features their pathologies are currently irreversible, and cognitive and behavioural problems are likely to increase over time, resulting in a need for long-term involvement of mental health services. Thus, an identifiable organic pathology does not necessarily classify a disease as 'neurological', and in many areas, such as dementia, learning disabilities, and neuropsychiatry, close collaboration between the specialties is needed to optimize diagnosis and management.

1.3 Main lines of neuroimaging research in psychiatry

Since the first reports of altered brain structures based on modern imaging techniques were published in the mid 1970s (Johnstone et al. 1976) neuroimaging and neurophysiology have established themselves amongst the most popular research methods in biological psychiatry. The direct, cross-sectional comparison between patients and appropriately matched control groups in the search for biomarkers of mental disorders is probably still the largest area of psychiatric neuroimaging and neurophysiology and will be covered in Chapters 6 and 7 of this book. The main interests driving this line of research are the search for markers that provide a more biological (and presumably sounder) foundation for psychiatric diagnoses and the quest for an understanding of the underlying biological mechanism, which may aid the development of more rational treatments. Increasingly, researchers are also investigating patients (or at-risk groups) longitudinally in order to identify prognostic markers that may, for example, improve the prediction of conversion from prodromal to clinical schizophrenia or mild cognitive impairment (MCI) to Alzheimer's disease (see Fig. 1.1).

The search for the biological mechanisms of mental disorders is enhanced by the investigation of carriers of genetic risk variants and

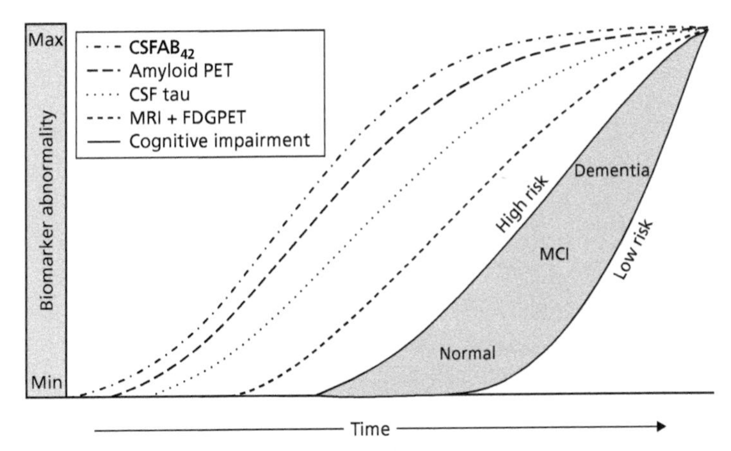

Fig. 1.1 Schema of the progression of biomarker changes preceding the onset of clinical symptoms of dementia. These include reduced amyloid-beta (AB_{42}) and increased tau protein in CSF, PET evidence for amyloid deposition in the brain, reduced glucose metabolism measured with FDG-PET, and brain volume reductions measured with MRI.

Reprinted from *The Lancet Neurology*, Clifford R Jack et al., 'Tracking pathophysiological processes in Alzheimer's disease: an updated hypothetical model of dynamic biomarkers', pp. 207–216., Copyright (2013) with permission from Elsevier.

the experimental (pharmacological) induction of psychiatric syndromes (Chapter 7). Another line of research concerns the investigation of treatment effects. This ranges from the demonstration of the mechanism of action of a particular drug, obtained through radioligand imaging, to the exploration of neuroplastic changes induced by long-term treatment and the search for biomarkers that can predict treatment success (and thus aid differential treatment decisions) or serve as surrogates for clinical outcomes in clinical trials (and thus make trials shorter and less expensive) (Chapter 9). Direct mapping of the neural correlates of the symptoms of mental disorders is mainly in the domain of functional magnetic resonance imaging (fMRI) and may provide a basis for inferring mental states from brain signals (Chapter 8). Once the neural correlates of a specific symptom or pathological mental state have been identified it may be possible to target the underlying neural networks with invasive or non-invasive physiological techniques. Chapter 10 will review some of the current trends in the development of such imaging-based or imaging-guided therapies.

1.4 Theoretical pitfalls of neuroimaging research

Most research in psychiatric neuroimaging and neurophysiology relies on group comparisons (between patients and controls, between different genetic groups, or between the same patients measured at different time points) because individual effects are too weak or unreliable. They are prone to the usual pitfalls of quantitative research, type I and type II errors (see Table 1.2), and to the publication bias often encountered in biomedical research. Type I errors occur if the null hypothesis (assuming no difference between groups) is rejected although it is true ('false positive result'). This can occur if statistical thresholds are not sufficiently stringent or if several statistical tests are conducted (implicitly or explicitly) and no correction for these multiple comparisons is applied. Type II errors occur if the null hypothesis is accepted although it is untrue ('false negative result'). This can happen if studies are underpowered, that is, if the sample is too small for the expected effect size. Although these are problems prevalent in any type of biomedical research, psychiatric imaging research is particularly vulnerable because of the large number of tests across brain regions and analysis protocols that are possible when analyzing neuroimaging data sets (risk of type I error) and because of the difficulties and costs involved in assembling sufficiently large samples (risk of type II error). The risk of type I error can be minimized by determining and registering analysis plans before the study, as is customary in clinical trials, and by replication in independent samples; several international consortia are addressing the problem of power by analyzing large multicentre data sets with standardized procedures (Shen et al. 2010, Stein et al. 2012, and Thompson et al. 2014).

Publication bias results from selective reporting of 'positive' findings (those that appear to refute the null hypothesis) and seems to affect psychiatry particularly strongly (Joober et al. 2012). One example

Table 1.2 Types of errors in statistical hypothesis testing

	H_0 is true	**H_0 is false**
H_0 is rejected	False positive (type I error)	True positive
H_0 is accepted	True negative	False negative (type II error)

H_0: "Null hypothesis", for example "There is no difference between the two groups".

where publication bias can have serious adverse effects on public health is the selective reporting of clinical trials. Because journals are generally more interested in publishing positive findings it will be difficult to abolish publication bias completely, and meta-analyses can ascertain and, to some extent, correct for this bias. Recent initiatives by some journals to solicit pre-registration of research protocols (Chambers 2015) and the requirement to register all trials on publicly accessible databases are important steps to reduce publication bias.

1.5 **Practical difficulties of psychiatric neuroimaging**

Most techniques of neuroimaging and neurophysiology require a great amount of cooperation on the part of the participant. Even lying still for several minutes, which is the minimal requirement for a high-resolution structural scan, can be a challenge for patients with mental or behavioural disorders, especially children. Head movement affects the signal quality and fidelity of localization for all neuroimaging procedures. Neurochemical imaging, both with MR-spectroscopy and radioligand imaging, integrates over long time periods (several minutes) and head movements thus affect the reliability of the biochemical information obtained. Finally, many functional imaging and cognitive neurophysiology paradigms require the participant to actively perform cognitive tasks, which require even higher levels of cooperation and attention and induce further sources of artefacts, for example, limb and eye movements. Even functional or metabolic imaging at rest or resting EEG/ Magnetoencephalography (MEG) can be heavily influenced by uncontrolled or uncontrollable differences in the mental activities of different participant groups. For example, differences in frontal glucose uptake, measured by positron emission tomography (PET), between patients with depression and control participants may be owed to a primary pathophysiological process in the frontal lobes or to different levels of rumination during half an hour of 'lying still' in a scanner—and these effects may even be intrinsically correlated and impossible to disentangle. Minimizing sources of noise and controlling or correcting for extraneous influences on the measured physical or physiological signals are thus particular challenges for studies with psychiatric patients. Although several techniques have been developed to account for the dosage of psychotropic medication none is fully satisfactory and thus present and past drug treatment are further inevitable confounders in most research studies with psychiatric patients, except for the very small proportion conducted on drug-naïve patients.

Claustrophobia is a problem for MRI studies (and other imaging modalities as well) and can lead to the premature termination of scans for some patients. Although claustrophobia can generally be overcome with appropriate anxiolytic medication, this may not be an option for research studies, and any additional medication can skew the results of functional studies. Anxiety reactions to the scanning environment appear to be more frequent in psychiatric patients than in the general population, which further complicates research studies aiming to recruit representative samples of the patient population. Researchers encounter particular challenges in studies of individuals with learning disabilities and dementia. A proportion of patients in these groups do not have capacity to consent to imaging studies, and those providing assent on their behalf will have to scrutinize the risk/benefit ratio for research studies particularly carefully, especially in the case of radioligand studies which entail potentially harmful exposure to ionizing radiation.

References

Chambers C. D. (2015) 'Ten reasons why journals must review manuscripts before results are known', *Addiction*, **110**(1):10–11.

Johnstone E. C., Crow T. J., Frith C. D., Husband J., Kreel L. (1976) 'Cerebral ventricular size and cognitive impairment in chronic schizophrenia', *Lancet*, 2(7992):924–6.

Joober R., Schmitz N., Annable L., Boksa P. (2012) 'Publication bias: what are the challenges and can they be overcome?', *J Psychiatry Neurosci.*, 37(3):149–52.

Shen L., Kim S., Risacher S. L., et al. (2010) 'Whole genome association study of brain-wide imaging phenotypes for identifying quantitative trait loci in MCI and AD: A study of the ADNI cohort', *Neuroimage*, **53**(3):1051–63.

Stein J. L., Medland S. E., Vasquez A. A., et al. (2012) 'Identification of common variants associated with human hippocampal and intracranial volumes', *Nat Genet.*, **44**(5):552–61.

Thompson P. M., Stein J. L., Medland S. E., et al. (2014) 'The ENIGMA Consortium: large-scale collaborative analyses of neuroimaging and genetic data', *Brain Imaging Behav.*, **8**(2):153–82.

Chapter 2

Techniques of neuroimaging: X-ray, CT, MRI, PET, and SPECT

Key points

- Magnetic resonance imaging (MRI) provides high resolution images of the human brain with exquisite tissue contrast
- Specific MRI techniques also enable imaging of fibre tracts (diffusion imaging), molecular profiles of brain tissue (MR-spectroscopy), and metabolism (functional imaging)
- Radioligand techniques enable neuropharmacological imaging; specific ligands are available for many neurotransmitter receptors
- Strategies for quantitative analysis of neuroimaging data can be divided into whole-brain and region-of-interest approaches
- Multivariate pattern analysis can be used for diagnostic classification, but it needs to be validated further before clinical applications can be contemplated

2.1 Radiography and CT

In medicine the term 'imaging' refers to techniques that produce images of the body or its parts. Specifically, it refers to a set of techniques that capitalize on the interaction between the body and an energy source and thus allow access to parts of the body that are not accessible by visual inspection. The recorded signals then allow for a reconstruction of two- or three-dimensional images of the respective body parts. Examples range from X-ray radiography to magnetic resonance imaging.

The electromagnetic spectrum and medical imaging

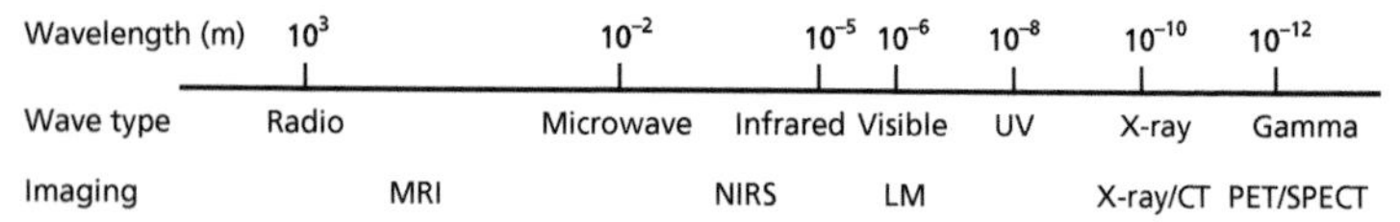

Fig. 2.1 The use of electromagnetic radiation of different wavelengths/frequencies in medical imaging. The radiofrequency spectrum is used for nuclear magnetic resonance (NMR)/MRI. Near infrared spectroscopy (NIRS) measures changes in blood oxygenation non-invasively through the skull using infrared light. Light microscopy (LM) uses waves from the visible spectrum. X-ray, computed tomography (CT), and the radionuclide imaging techniques SPECT and PET utilize ionizing radiation from the high-frequency (and high energy) end of the spectrum.

Conventional medical radiography using high energy electromagnetic radiation (X-rays) was invented by the German physicist Wilhelm Conrad Röntgen in 1895 (see Fig. 2.1). It is based on the principle that different tissues absorb radiation to a different extent. The image is created by shining a beam of X-rays through the relevant part of the body and placing a detector behind it. Areas of high absorption, for example, bones, will appear bright on the detector whereas areas of low absorption, for example, air-filled spaces like the lungs, will appear dark. For many decades, photographic film was used as the detector, but present-day radiography is mostly digital. Whereas conventional radiography produces two-dimensional images of superimposed body structures, three-dimensional reconstruction can be obtained by tomographic procedures, which entail moving the radiation source and the detector around the body part of interest. Computer-assisted tomography (CAT or CT) provides virtual cuts through the body based on the signals created by an X-ray source that moves around the body in the axial plane. Modern CT scanners can generate detailed images of the brain with a typical slice thickness of 5mm in only a few minutes.

The difference in the natural absorption of electromagnetic radiation between tissues of interest is often rather low. For example, it is difficult to distinguish blood vessels from brain tissue on CT images. However, the radioabsorption of blood vessels can be increased by injecting substances that have a higher average atomic number than blood because the attenuation of the X-ray beam is affected by the number of electrons within the tissue. Iodine, the most commonly used contrast medium, has a much higher atomic number than hydrogen or oxygen, the main constituents of blood. Thus, blood vessels will appear bright on a contrast-enhanced CT scan.

CT of the head, without or with contrast medium, has many indications in neurology and traumatology, including acute head trauma, suspected intracranial haemorrhage, other vascular lesions, and cranial malformations (American Society of Neuroradiology 2014). The use of an intravenous or intrathecal (injected into the subarachnoid space) contrast medium improves the demarcation of vascular or CSF spaces. Because most brain lesions can lead to altered mental states, psychiatrists, especially liaison psychiatrists, will often encounter patients for whom a CT of the head has been requested. However, for most differential diagnostic scenarios in psychiatry where imaging is required, magnetic resonance imaging (MRI) will be the method of choice (see Chapter 4). CT will still be needed where MRI is not available, or for patients who cannot undergo MRI because of safety considerations, for example, owing to cardiac pacemakers or ferromagnetic implants.

2.2 Magnetic resonance imaging (MRI)

2.2.1 Principles of MRI

Magnetic resonance imaging (MRI) was introduced into diagnostic radiology in the 1970s. It uses the phenomenon of nuclear magnetic resonance displayed by atomic nuclei with an odd number of protons and/or neutrons in a magnetic field. These nuclei have a net nuclear spin and thus behave like tiny magnets themselves. They absorb and re-emit electromagnetic radiation at a specific frequency that is proportional to the strength of the static magnetic field. Let us consider the example of hydrogen (1H), the nucleus most often used for MRI, which is made up of just one proton. In a static magnetic field of 1.5 Tesla, the field strength of standard clinical MR systems, its resonance frequency is around 64 Megahertz (MHz), which happens to be in the radio-frequency (RF) part of the electromagnetic spectrum. This is the frequency at which the protons precess around the axis of the static magnetic field (see Fig. 2.2a). The difference in the number of spins aligned parallel to the static magnetic field and those aligned antiparallel to it gives rise to a net magnetization (Fig. 2.2b). The net magnetization can be disturbed by applying a RF pulse at the resonance frequency, which rotates the magnetization away from its equilibrium alignment (Fig. 2.2c). The amount of rotation of the net magnetization vector is called a 'flip angle'. The change in electromagnetic signal produced by the rotation of the magnetization vector can be picked up by a receiver coil (antenna) because changing magnetic fields induce electric currents, according to Faraday's Law of Induction.

The signal picked up by the receiver coil is influenced by the time it takes for the proton spins to re-align with the static magnetic field (spin-lattice or longitudinal magnetic relaxation, described with the time constant T1) after the RF pulse is switched off. In addition, each spin influences the magnetic field of the other spins around it; each spin will precess with a slightly different frequency and thus lose its coherence with the other spins, resulting in reduction of the net magnetization. This dephasing is described by the time constant T2 ('transverse relaxation time'). Additional local differences in the magnetic field lead to even more rapid dephasing and a shorter T2, described by the effective transverse relaxation time T2*.

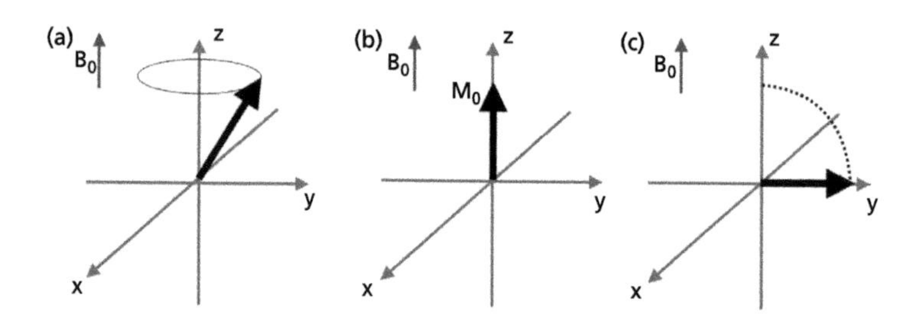

(a) In the presence of a magnetic field, B_0, spins precess about the axis of the applied field (z), at a characteristic frequency known as the Larmor Frequency.

(b) The ensemble of spins produce a net magnetization, M_0, aligned with the B_0 field.

(c) An excitation RF pulse, at the Larmor frequency, tips the magnetization into the x-y plane, where it can be detected.

Fig. 2.2 Longitudinal and transverse magnetization.

Courtesy of Dr John Evans.

Table 2.1 Approximate values for the time constants for recovery of longitudinal magnetization (T1) and decay of transverse magnetization (T2) at 1.5 Tesla. T1 values increase with field strength of the static magnetic field whereas T2 values are not affected very much by field strength.

	Grey matter	White matter	CSF	Fat	Blood
T1	900ms	600ms	4000ms	250ms	1200ms
T2	100ms	80ms	500ms	110ms	360ms

T1 and T2 are influenced by the molecular environment of the protons, and thus ultimately by the tissue type. Table 2.1 provides examples of approximate T1 and T2 values at 1.5 Tesla. T2* is always shorter than T2, for example, around 65ms in grey matter at 1.5 Tesla.

Tissues with similar T1 or T2 are difficult to distinguish. Conversely, the image contrast becomes better with increasing difference in these time constants across the tissues of interest. For example, the contrast between white and grey matter on T2-weighted MRI scans is poor, whereas that between white matter and CSF is excellent (see Fig. 2.3). Scans can be 'weighted' for a particular contrast, for example, T1 or T2, by choosing appropriate parameters of the MR sequence that determine the time points of RF pulse generation (TR) and signal reception (TE).

The procedures described so far provide image contrast between tissue types, but not localization. Information about the spatial source of a signal, which allows for the reconstruction of detailed three-dimensional images, is obtained through spatial variations of the strength of the static magnetic field. Through additional gradients in the three Cartesian directions, each volume element (voxel) in the imaged cube can be characterized by a slightly different resonance frequency, which enables the 'spatial encoding' of the acquired MR signal (see Fig. 2.4).

2.2.2 Magnetic resonance spectroscopy

The MR signal can also be used to obtain information about the chemical composition of a particular voxel. This technique, magnetic resonance spectroscopy (MRS), also mainly uses the proton (^{1}H) signal, although in-vivo MRS with other nuclei, particularly ^{31}P (phosphorus), is also possible. ^{1}H-MRS utilizes the slight differences

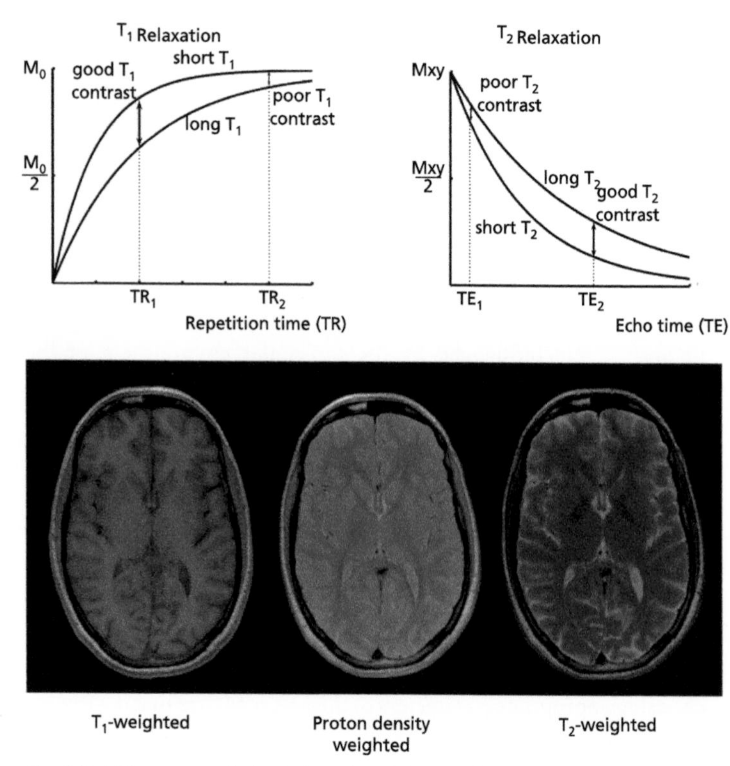

Fig. 2.3 MR contrasts in brain imaging. The graphs in the upper row describe the relationship between different sequence parameters (TR, TE) and optimal T1 and T2 contrast. The images in the lower row provide examples of the brain tissue contrasts obtained with T1-, PD-, and T2-weighted MRI. The following sequence parameters were used.

T1: fast spoiled gradient echo; TR = 8ms; TE = 3ms; Inversion Time = 450ms; flip angle = 20°.

PD: fast spin echo; TR = 3000ms; TE = 9ms; flip angles = 90°; 180°.

T2: fast spin echo; TR = 3000ms; TE = 81ms; flip angles = 90°; 180°.

Courtesy of Dr John Evans.

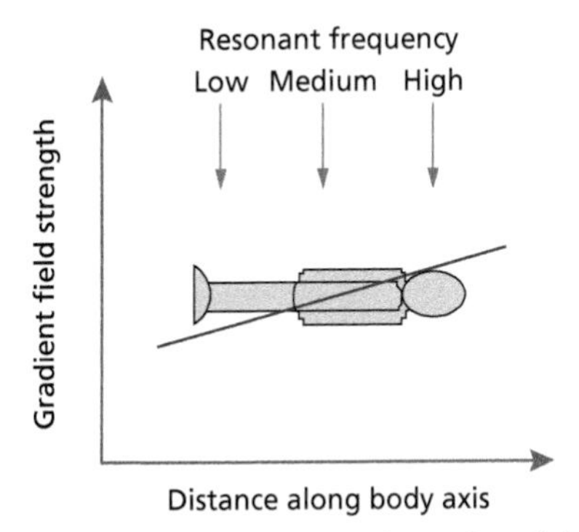

Fig. 2.4 Principles of spatial encoding through gradient fields. The strength of the gradient field varies along the body axis, resulting in slight changes in the local magnetic fields and spatially specific resonance frequencies. This allows for spatial encoding of the MR signal in the direction of the body axis. Extension of this technique to all directions of the Cartesian coordinate system allows for three-dimensional spatial encoding.

in the resonance frequencies of protons in different molecules (called 'chemical shift'), which are owed to the interaction of the nuclear spins within the molecule with the large magnetic field generated by the scanner. These differences are expressed in parts per million (ppm) as a fraction of the main magnetic field. Because of the abundance of water in the body, additional steps need to be taken to suppress the signal arising from water. Failure to do so will lead to artefacts in the baseline of the spectrum and hamper quantification.

Each metabolite has a characteristic 'fingerprint' of resonance peaks, and the area under these peaks reflects the local concentration of the metabolite (see Fig. 2.5). By fitting the known set of resonance peaks associated with each metabolite to the spectrum measured in vivo, one can generate semi-quantitative information about metabolite concentrations. Because of the low signal to noise of MRS, several minutes of data acquisition and large voxels are needed for a good spectrum for a single voxel (commonly between 1 and 27 cubic centimetres).

Because several of the metabolites that can be resolved through MRS at standard field strengths, for example, N-acetylaspartate (NAA), myo-inositol, and choline, reflect cell metabolism and division, MRS is being developed as a tool to aid the differential diagnosis

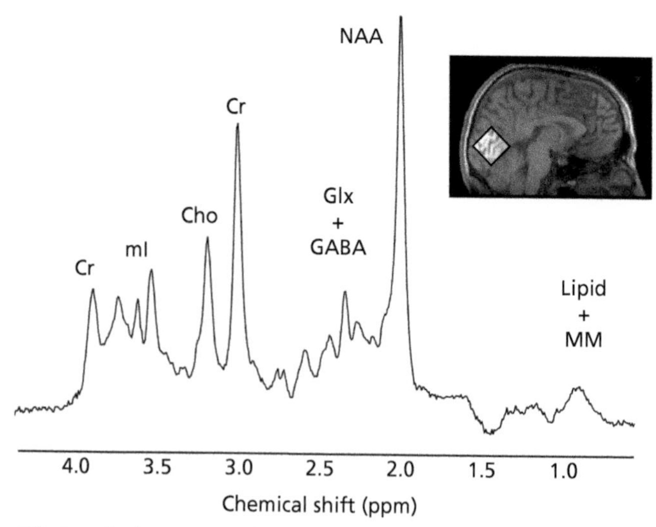

Fig. 2.5 Standard spectrum identified by 1H-MRS at 3 Tesla from the human brain. (See inset for occipital lobe.) Cho = Choline; Cr = Creatine; Glx = Glutamate/ Glutamine; mI = Myo-Inositol; MM = Macromolecules; NAA = N-Acetylaspartate.

Courtesy of Dr John Evans.

of brain tumours (Ryken et al. 2014). MRS has also been established as a research tool in neurology and psychiatry because of its capacity to measure markers of neuronal integrity (NAA) and neurotransmitters (GABA, glutamate/glutamine) (Wijtenburg et al. 2015 and Spencer 2014). The increasing availability of 7 Tesla MR systems for research is likely to boost this field further because higher field strengths allow for finer differentiation of molecular spectra. As explained earlier in this section, the chemical shift is measured in parts per million of the main magnetic field and thus increasing the main field increases the discriminability between peaks for individual substances, for example, enabling a distinction between glutamate and glutamine, which is not easily obtainable at lower field strengths.

2.2.3 **Diffusion imaging**

Diffusion-weighted imaging (DWI) produces images that capture differences in the diffusion of water across tissues. Whereas water molecules move freely in CSF (isotropic diffusion) their diffusion is hindered (anisotropic) in white matter, mainly through cell membranes and to a much smaller extent through myelin sheaths. Although in principle this applies to cortical tissue as well grey matters is typically considered isotropic at the spatial resolution obtained by standard imaging protocols. Pulse sequences are generally sensitive to the effects of diffusion because the incoherent motion of water molecules within the presence of a gradient will lead to extra signal dephasing, beyond that of T2/T2*. As most sequences utilize gradients, there will always be some diffusion weighting (e.g. from the slice-select gradients), but the impact will be small. To enhance this, diffusion-weighted scans typically introduce significant gradient amplitudes that amplify the impact of the intravoxel incoherent motion. The resulting diffusion-weighted images are clinically useful in the detection of early ischaemic brain damage and traumatic and hypoxic brain injury. In order to compute the average diffusivity in a voxel, it is sufficient to apply DWI pulses in three orthogonal directions. Rapid estimates of the mean diffusivity of brain tissue can even be obtained with a single pulse (Mori and van Zijl 1995 and Wong et al. 1995), which may be attractive for clinical applications in emergency settings.

DWI with at least six (but today, more commonly 30 or 60 directions) can also be used to reconstruct the orientation of fibre pathways through the white matter of the brain. It allows one to determine the predominant orientation of fibre tracts in a voxel through the computation of the diffusion tensor, which can be illustrated by fitting an ellipsoid to the diffusion data for each voxel (see Fig. 2.6). The estimation of the diffusion tensor requires at least six directions. With 30 or more directions it becomes possible to resolve more than one fibre orientation within each voxel.

The tensor information for each voxel can be used to obtain more quantitative data on local diffusivity (fractional anisotropy, axial and radial diffusivity) (see Fig. 2.6). These diffusion tensor imaging (DTI) data can also provide the basis for the reconstruction of fibre tracts in the brain and spinal cord (see Fig. 2.7). However, the tensor model is limited in its ability to reconstruct pathways in that it only resolves one orientation within a voxel and therefore cannot deal appropriately with crossing fibre regions, of which there are many in the brain.

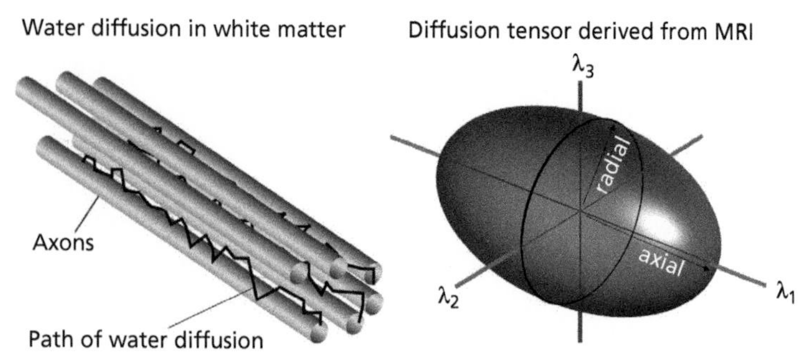

Fig. 2.6 Principles of diffusion tensor imaging. The left panel demonstrates the restriction imposed on water diffusion by axons (and by the myelin sheaths surrounding the axons) in fibre tracts. The right panel visualizes the diffusion tensor likely to be seen in this structure (with the eigenvalues λ1, λ2, and λ3). The main parameters obtained from diffusion imaging are axial (λ1), radial ([λ2 + λ3]/2), and mean diffusivity ([λ1+ λ2+ λ3]/3) and fractional anisotropy (FA) (which can also be calculated from the eigenvalues of the diffusion tensor). FA can range from zero (completely isotropic diffusion, that is, all eigenvalues are equal) to one (completely anisotropic diffusion, where there is only one non-zero eigenvalue).

Courtesy of Dr Mark Drakesmith.

2.2.4 Functional imaging

Imaging becomes 'functional' when the signal that is picked up by the technique is not just a static contrast between different tissue types (as in anatomical or structural imaging), but dependent on blood flow or chemical or electrical processes in the tissue. In the case of the brain, these functional processes picked up by functional neuroimaging are mainly the metabolism of glucose and oxygen or the differences in the binding of neurotransmitters and other signalling molecules. Three principal MRI techniques can pick up such functional signals: perfusion-weighted imaging (PWI) using a contrast agent, functional magnetic resonance imaging (fMRI), and arterial spin labelling (ASL), a perfusion-weighted technique that uses blood as endogenous contrast.

PWI utilizes a paramagnetic contrast agent, most commonly chelated complexes of the rare earth gadolinium. Paramagnetic materials are weakly attracted by magnetic fields and create local inhomogeneities

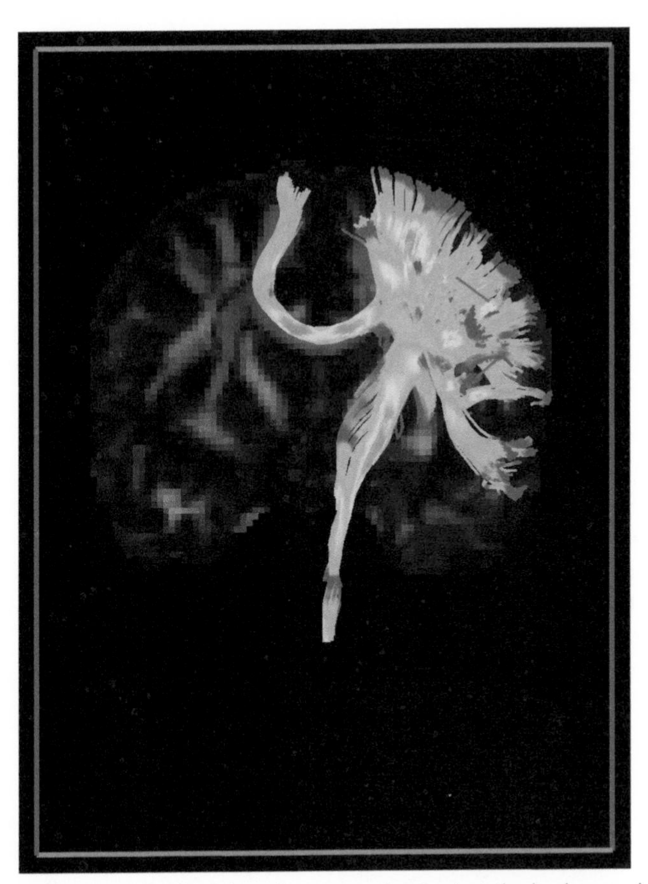

Fig. 2.7 An example of DTI-based tractography. Against the background of a tensor image, in which the diffusivity in each of the Cartesian directions is colour coded, we see the corpus callosum (connecting the two hemispheres) and the corticospinal tract. The image was obtained using a seed region in the left motor cortex (right side of the image) and shows how it connects to the motor cortex of the other hemisphere through the corpus callosum, to the spinal cord through the corticospinal tract, and also to other parts of the motor cortex in the same hemisphere. The tractography was performed with the software ExploreDTI (Leemans et al. 2009).
Courtesy of Dr Mark Drakesmith.

in the static magnetic field, resulting in reduced T2* and signal loss. PWI of the brain is commonly based on the technique of dynamic susceptibility contrast (DSC)-MRI, which involves the rapid intravenous injection of the contrast agent and the serial measurement of the signal loss in T2- or T2*-weighted images, produced as this bolus of paramagnetic material travels through the vessel. Because the passage of an intravenously injected bolus of gadolinium-based contrast agent is fast, data need to be acquired in a few seconds, and the resulting images trace the distribution of blood in the brain. Because PWI can thus detect areas of reduced blood supply before any resulting damage to brain tissue that would show up on other MRI sequences, for example, DWI, it is particularly useful in the early diagnosis of stroke and for determining the risk of future lesion enlargement (Neumann-Haefelin et al. 1999). Other uses for contrast-enhanced MRI include the improved delineation of vessels for MR-angiography and the demonstration of disruptions of the blood brain barrier. The blood brain barrier, which is formed by the endothelial cells of the capillaries and surrounding astrocytes and separates the blood from the extracellular fluid of the brain, is generally impermeable for gadolinium-based contrast agents. Thus, enhancement of lesions by gadolinium (for example, certain tumours or areas of acute inflammation) reflects a breakdown of the blood brain barrier.

FMRI provides an indirect measure of neural activity without the need for an external contrast agent. Synaptic activity leads to local vasodilatation through the release of vasoactive substances and through direct control of vessel smooth muscle by astrocytes and neurons. The ensuing influx of fresh blood provides more oxygen than is required for the increased aerobic glycolysis needed to restore membrane potentials (ATPase activity is energy dependent (Linden 2012a)). The vasodilation triggered by neural activity thus leads to local oversupply of oxygen and shifts the ratio of oxygenated (oxy-) to deoxygenated (deoxy-) haemoglobin in favour of oxy-haemoglobin. Because deoxy-haemoglobin is paramagnetic, the result is an increase of T2*-weighted signal (essentially, the opposite of the effect of gadolinium described above). The net effect of inflow of fresh, oxygenated blood into a part of the brain on the MR signal is thus positive, a signal increase. This component of the MR signal is called blood oxygenation-level dependent (BOLD) signal, and has over the last 20 years become a very important surrogate marker of neural activity in both basic and clinical imaging research. In colour-coded statistical maps, this increase of the BOLD signal will normally be denoted in warm colours. For example, when we measure brain activity in patients experiencing auditory hallucinations with fMRI (see Chapter 8), we can detect a hotspot in their auditory cortex (even without any change in external auditory stimulation). The technique of echo-planar imaging (EPI) allows for the acquisition of whole-brain functional images at a rate of two seconds or less, which is a reasonable temporal resolution for many mental processes, but still three orders of magnitude below that of EEG and MEG (see Chapter 3). A further limitation of the temporal resolution of fMRI arises because of the 'haemodynamic delay'—it takes about five seconds from the onset of the neural activity for the resulting BOLD effect to emerge (see Fig. 2.8).

Another MRI technique to map blood distribution is ASL. Unlike PWI, ASL does not need an externally administered contrast agent but 'labels' blood before it enters the brain by a radiofrequency pulse that inverts hydrogen spins in its water molecules (Jahng et al. 2014). The labelled spins then travel through the brain, and a proportion of them exchange into the brain tissue and stay there. After the spins

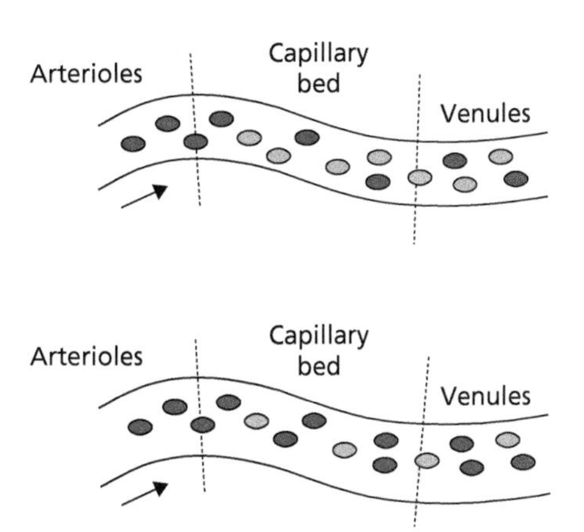

Fig. 2.8 The BOLD effect of fMRI. Upper panel: distribution of oxygenated (dark grey) and deoxygenated haemoglobin (light grey) during a baseline state. Lower panel: during mental activity or external sensory stimulation fresh blood flows through dilated vessels into the specific brain region, increasing the ratio of oxygenated versus deoxygenated haemoglobin molecules. This leads to an increase of the MR signal measured with sequences sensitive to the effective transverse relaxation time T2*.

Courtesy of Professor Peter Jezzard.

that remain in the blood have been washed out, a scan is acquired and subtracted from a reference image obtained without labelling. The resulting difference map indicates the regional cerebral blood flow. ASL techniques are used in the investigation of cerebrovascular and neurodegenerative disorders and can provide maps of activated brain regions during perceptual and cognitive tasks and after the administration of drugs (Detre et al. 2009 and Detre et al. 2012). Initial studies have also applied ASL to psychiatric questions, for example, investigating alterations in temporal lobe blood flow in patients with auditory verbal hallucinations (Homan et al. 2013 and Petcharunpaisan et al. 2010). However, its temporal resolution and signal-to-noise ratio are inferior to those of BOLD fMRI (Borogovac et al. 2012). It would be attractive to combine the strengths of both techniques, using ASL to calibrate BOLD fMRI. This approach entails the acquisition of an ASL scan during administration of carbon dioxide, which induces hypercapnia (increased carbon dioxide in the blood) and vasodilation. ASL scans acquired during hypercapnia and hyperoxia (induced through oxygen breathing) provide the data that allow for an estimation of the local cerebral metabolic rate of oxygen ($CMRO_2$) and the oxygen extraction fraction as well as cerebral blood flow (CBF) and cerebral blood volume (CBV) (Wise et al. 2013). With such calibration it becomes feasible to estimate $CMRO_2$ from simultaneously acquired BOLD and ASL data and thus obtain quantitative (rather than purely relative) information from fMRI data. This is particularly important for comparisons between groups or within-group comparisons at different measurement points (for example, before and after a treatment) because any changes in relative activation could be caused by changes in tonic baseline activity as much as by the phasic activity induced by the experiment of the respective study. For example, an increase in the BOLD signal change in the amygdala during the presentation of affective pictures after a treatment could indicate a genuine increase in amygdala reactivity, but also a change in baseline metabolic activity, or a combination of both. On the basis of calibrated fMRI it is possible to distinguish between these possibilities.

2.3 Radioligand imaging

The first wave of functional neuroimaging techniques for use in humans, developed and refined during the 1960s to 1980s, relied on radioactivity. By injecting patients or research participants with molecules that were radioactively labelled one could trace the distribution of these molecules in the body by picking up the radioactivity emitted as these radioactive markers decayed. Several scanning techniques have been developed for this purpose, including single photon emission computed tomography (SPECT) and positron emission tomography (PET). They are still very important for the investigation of abnormalities in brain chemistry in patients with neurological and psychiatric disorders and to ascertain that drugs used for these disorders act in humans in the way predicted from animal studies (Linden 2012b). However, their use to track changes in neural activity during specific mental processes (by looking at changes in glucose or oxygen consumption) has largely been superseded by the non-radioactive technique of BOLD-fMRI (see Section 2.2.4).

PET and SPECT are based on the detection of the products of radioactive decay from radioactively labelled tracers. These are produced by adding a radioactive isotope, for example, ^{18}F (fluorine) for PET or ^{123}I (iodine) for SPECT, to a biologically active molecule. For example, ^{18}F can be added to deoxyglucose to yield fluoro-deoxyglucose (FDG), which crosses the blood brain barrier and is taken up into brain cells. However, unlike glucose, it is not metabolized further. FDG uptake thus provides a good measure of regional glucose utilization. ^{18}F is unstable, and one of its protons is converted into a neutron, yielding ^{18}O (oxygen). During this conversion of a proton into a neutron, a neutrino and a positron are emitted. The positron travels for a few millimetres and then reacts with an electron (annihilation) to yield two photons, which move in opposite directions and can be detected by a special camera (a PET scanner). The camera only counts photons that arrive at opposite ends at about the same time (coincidence detection). Because many such annihilation events will happen at about the same time and lead to random directions of gamma rays, the position of the emitting positron can be calculated as the crossing point of these lines (Fig. 2.9) with a spatial resolution in the millimetre range.

Isotopes used for SPECT emit a single photon, which can be detected by a gamma camera. The position of the emitter is determined through a device called a collimator, a block of lead with holes that is interposed between the scanned object (e.g. the brain) and the crystal that detects

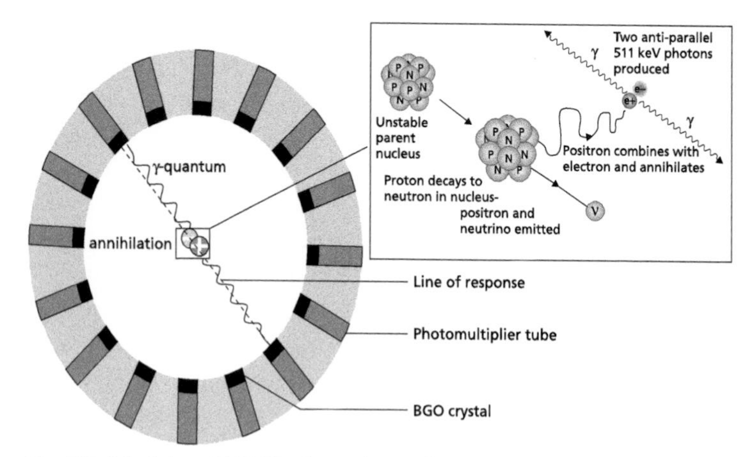

Fig. 2.9 Principles of PET. The inset shows the radioactive decay of the unstable parent nucleus through the conversion of a proton (P) to a neutron (N), emitting a neutrino (Greek letter ν, nu) and a positron. When the positron (e+) encounters an electron (e–), they annihilate, resulting in two photons (Greek letter gamma, γ), which can be picked up by the crystals in the detector ring (large picture). BGO: Bismuth germanium oxide, a scintillator crystal used in PET systems.

Reproduced from David Linden, *The Biology of Psychological Disorders*, 2011, with permission of Palgrave Macmillan.

the photons. The collimator blocks photons from reaching the detector crystal, unless their direction of travel is perpendicular to the plane of the crystal surface. It thus allows for a linear reconstruction of the source of the emission. For three-dimensional images, the gamma camera has to be moved around the object, resulting in longer scan times than in PET. The spatial resolution is also lower, but SPECT is cheaper and more widely available. Its radionuclides have a longer half-life than those commonly used in PET. The effective radiation doses for both techniques are in a similar range, for example, 8 millisievert (mSv) for a standard FDG-PET scan and 4mSv for a SPECT-based DaTSCAN (TM), which uses a ligand for the dopamine transporter.

PET and SPECT are the main methods for the study of neuropharmacological effects in the living human brain. There is ample literature on the dopamine, serotonin, acetylcholine, GABA, glutamate, and opioid systems (Nikolaus et al. 2009). Synaptic levels of neurotransmitters can only be measured indirectly through changes in receptor occupancy. For the dopamine system, several D_2 receptor radioligands, including ^{11}C-raclopride and ^{18}F-fallypride for PET and ^{123}I-IBZM (iodobenzamide) for SPECT are available. According to the standard occupancy model, higher ligand binding reflects lower synaptic neurotransmitter levels and vice versa (Laruelle 2000). At the presynaptic level, dopamine synthesis can be assessed by measuring DOPA decarboxylase activity, and dopamine storage can be assessed through the uptake of ^{18}F-DOPA. Ligands of the dopamine transporter (DAT) can be used to demonstrate target engagement for DAT inhibitors, which include the main class of drugs used for the treatment of attention deficit hyperactivity disorder (ADHD) (e.g. methylphenidate) and several drugs of abuse (Nutt and Nestor 2013). A SPECT ligand for the DAT is also being used to probe loss of dopaminergic neurons in Parkinson's disease (DaTSCAN (TM); see Chapter 4).

2.4 **Analysis methods**

Most clinical interpretation of neuroimaging or neurophysiology data is based on the visual inspection of the images or traces, with the main exception of sleep-EEG studies where automated analysis is adduced to determine sleep stages. However, research in psychiatric imaging and neurophysiology is largely based on the comparison of group data (patient and control groups or longitudinal comparison of clinical cohorts) that requires statistical analysis. This section therefore explains the principles of common analysis methods in psychiatric imaging, without aiming to be exhaustive. Readers wishing to learn more about the state of the art of data analysis are referred to specialized textbooks or review papers on fMRI (Poldrack et al. 2011, Huettel et al. 2014, and Friston 2007), diffusion MRI (Jones 2011), structural imaging (Mills and Tamnes 2014), EEG (Kamel and Malik 2015), MEG (Supek and Aine 2014), and radionuclide imaging (Seeman and Madras 2014).

2.4.1 **Structural imaging**

The normal volume of the adult human brain is generally between 1 and 1.5 litres. Thus, a structural MRI scan covering the whole brain with the standard resolution of 1mm × 1mm × 1mm (= 1 microlitre) contains over one million volume elements (voxels), each of which could potentially contain important information about differences between patients and controls. Running separate statistical tests on all these voxels would make it challenging to apply the required correction for a multiple comparison (or require huge participant numbers that can rarely be achieved in clinical imaging studies). Furthermore, because all brains vary slightly in their exact morphology it is challenging to merge imaging data across participants. Two principal strategies have been developed to address these challenges: template-based anatomical alignment and region-of-interest (ROI) analyses (Fig. 2.10).

The most widely based analysis technique using template-based alignment is voxel-based morphometry (VBM) (Ashburner and Friston 2000). VBM involves the normalization of the structural scans of all participants of a study into a stereotactic reference space, or onto a study-specific template brain. The different tissue

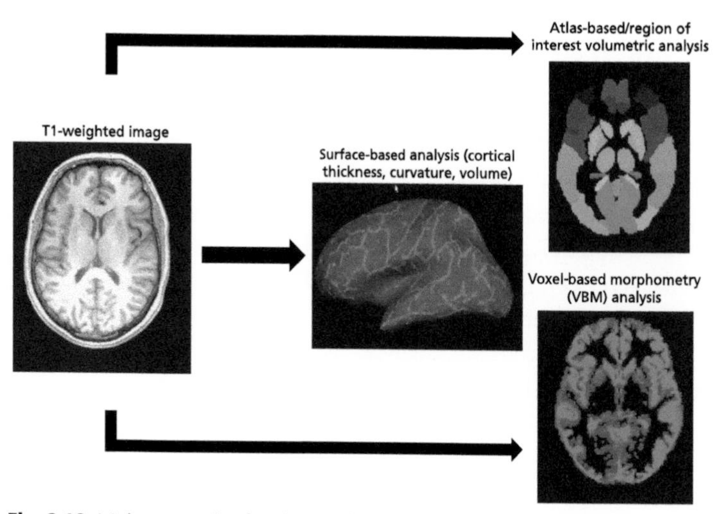

Fig. 2.10 Main strategies for the analysis of structural MRI data. A high-resolution T1-weighted image set of the whole brain can be parcellated into regions of interest (for example, with automated anatomical labelling (Tzourio-Mazoyer et al. 2002)) or analyzed with voxel-based morphometry (VBM) (see Fig. 2.11). Surface-based analyses entail individual segmentation of the cortical sheath and give access to more specific measures of cortical thickness and curvatures. Cortical reconstruction was performed with the Freesurfer image analysis suite (http://surfer.nmr.mgh.harvard.edu/) (Dale et al. 1999).

Courtesy of Dr Mark Drakesmith.

compartments (grey, white matter, CSF) are then automatically segmented based on their probability values. This technique allows for a voxel-wise comparison of grey matter density, and spatial smoothing is used to control for subject-specific misalignments that can occur during the warping of the template (Fig. 2.11). VBM is an appropriate tool when a researcher is agnostic about the specific brain regions associated with the independent variable, for example, the comparison between patients and controls. It relies on a mathematical technique called random field theory to address the multiple comparison problem. Other approaches that have been proposed to correct for multiple comparisons in structural and fMRI data include the Bonferroni correction, false discovery rate (FDR), and permutation methods, which are explained in detail in Poldrack et al. (2011), Huettel et al. (2014), and Friston et al. (2007). Whereas VBM normalizes brains in three-dimensional space, another approach aligns brains according to their surface structure to take account of individual differences in cortical folding patterns (Goebel et al. 2006 and Dale et al. 1999).

ROI analyses are based on measurements of pre-specified anatomical areas, for example, the hippocampus, cingulate gyrus, particular subfields of the frontal lobe or basal ganglia nuclei. These measurements of the grey matter volume (or thickness in the case of cortical regions) of a particular region of the brain can be obtained through manual segmentation—still considered to be the gold standard of quantitative structural imaging—or automatic segmentation and parcellation procedures, which show good concordance with manual segmentation, at least in young adults (Morey et al. 2009). Automatic segmentation of grey matter boundaries and parcellation into anatomical regions based on probabilistic atlases is available in several software packages, but users need to have a good knowledge of neuroanatomy in order to ensure the relevant quality control. The resulting volumetric measures can then be entered in statistical tests for group comparisons or into regression analyses probing for associations between continuous variables, for example, personality traits, and regional brain volume. Segmentation and parcellation into a priori-defined regions is effective if researchers have strong hypotheses about specific brain regions.

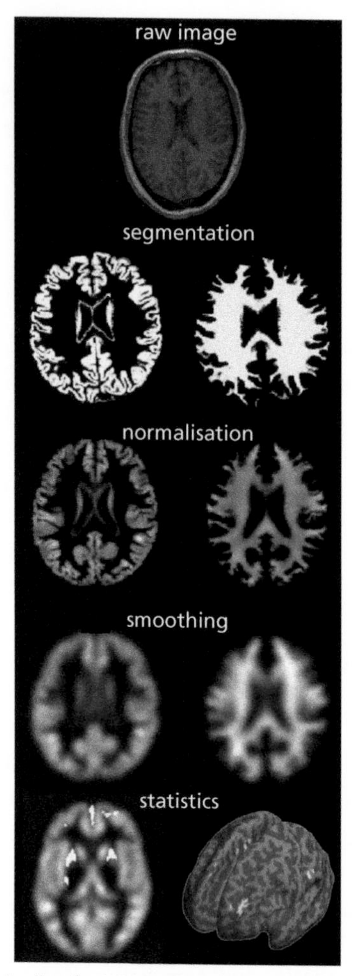

Fig. 2.11 Analysis pipeline for VBM using the software package Statistical Parametric Mapping (SPM) 8 (http://www.fil.ion.ucl.ac.uk/spm/software/spm8/). This workflow starts with the automatic segmentation of grey (left column) and white matter. After normalization and smoothing of the resulting images, group comparison of density/volume of grey or white matter is possible, and statistical differences can be colour coded.

Courtesy of Dr Thomas Lancaster.

These tools are often used in multi-centre studies because the output values can be subjected to meta-analyses, avoiding the need to run huge group analyses on the primary imaging data. Compared to VBM, ROI analyses have lower sensitivity to detect small intra-region effects.

2.4.2 Diffusion imaging

By comparing signal intensities for different (at least two) b-values DWI produces voxel-wise maps of diffusivity, measured by the apparent diffusion coefficient (ADC). Although the ADC can be estimated reasonably well with three image directions, more directions (theoretically, at least six) are needed for the modelling of fibre directions and the computation of more detailed information about tissue diffusivity. Measurements from six or more directions allow for the computation of the diffusion tensor, which can be visualized as an ellipsoid (see Fig. 2.6) and yield the parameters of fractional anisotropy (FA), mean diffusivity (MD), axial diffusivity (AD), and radial diffusivity (RD) for each voxel. AD describes the diffusion along the principal axis of the tensor, RD the diffusion perpendicular to this axis, and MD the average diffusivity in all directions. These values can be summarized by the FA, which is maximal if AD is large compared to RD (indicating one principal diffusion direction as in a single population of straight axons) and minimal if water can diffuse evenly in all directions as in CSF. FA values range from 0 to 1 and typical values for white matter are between 0.3 and 0.6.

FA values can reflect the integrity of bundles of axons, and thus, comparisons of FA maps between patient groups and controls may provide insight into potential alterations of white matter microstructure in mental disorders, which are often conceptualized as disorders of connectivity. Such comparisons can be performed at a whole-brain level, although this approach will run into similar problems with the sheer amount of statistical tests performed as other whole-brain analyses. Furthermore, it is vulnerable to partial volume effects. If two groups differ in size of their lateral ventricles, voxels that are anatomically matched by the normalization procedure may systematically contain more CSF in one group than another. This would lead to lower FA values, however, these may not necessarily reflect any change in the actual fibre architecture.

One attractive way of data reduction which also reduces the partial volume effect problem restricts the analysis to a skeleton of the main fibre tracts of the brain. Using this tract-based spatial statistics (TBSS) analysis, implemented in the FSL software (Smith et al. 2006), investigators can be reasonably certain that any differences in

FA actually arise from white matter unless gross anatomical distortions are present. Furthermore, TBSS increases statistical power compared to whole brain analysis (WBA) by reducing the variance of the diffusion parameters of interest. Even better anatomical specificity can be obtained with tractographic analyses, which trace individual fibre tracts from anatomical seed regions using an algorithm that follows the preferred direction of diffusion. For example, by using a seed region in the motor cortex and the ipsilateral anterior pons one can trace the corticospinal tract based on the diffusion tensor model (Fig. 2.7). This approach then allows researchers to extract mean FA and diffusivity values, or even values from a separate protocol (e.g. myelination (Deoni et al. 2008)) for an entire fibre tract. The utility of tractography is not confined to group studies because it can be useful in mapping out alterations of fibre paths caused by brain lesions for pre-operative mapping. Because the standard tensor model is based on the assumption of a single main diffusion direction and thus cannot properly estimate the directions of crossing fibres within a voxel, other mathematical approaches have been proposed that utilize high angular resolution diffusion-weighted imaging (HARDI), generally with over 30 diffusion directions, and allow for a fuller, more accurate, reconstruction of the actual white matter anatomy (Abhinav et al. 2014).

2.4.3 **FMRI**

Some of the principles of fMRI analysis can be illustrated through a standard functional brain mapping experiment. If the experimenter alternates the presentation of houses and faces on the projection screen during the fMRI session, he/she will be able to compute the differences in the (hypothesized) hemodynamic response at each voxel across the whole brain. In order to compute such contrasts one has to generate a regression model of the experiment that takes into account the periods when one or the other kind of stimuli was presented and the haemodynamic delay, and apply this model to the data. The differences in the haemodynamic response between these conditions at each voxel in the brain can be summarized in a contrast map that highlights (with colour-coded statistical values) areas where presentation of one set of stimuli (for example, faces) elicits a greater response (see Fig. 2.12).

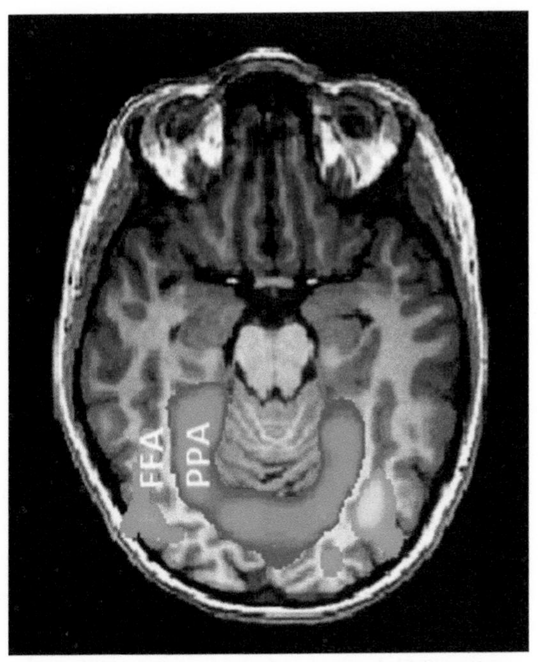

Fig. 2.12 Example of a fMRI activation map. The image shows areas in the occipital and temporal lobes whose activation differs during the presentation of faces (warm colours, including the fusiform face area ((FFA)) and houses (cold colours: parahippocampal place area (PPA)).

Contrasts between conditions and groups (and the interaction of these factors) can be evaluated through a comparison of these beta weights using analysis of variance (ANOVA). As with WBA of structural imaging data, multiple comparison correction is a major challenge. One way of minimizing this challenge is to restrict the analysis to pre-specified and independently localized ROIs. For example, an investigator wishing to find out whether house or face areas are more responsive to pictures of animals could localize the main house-selective area, the PPA, and the main face-selective area, the FFA, with a localizer scan (see Fig. 2.12) and then compare the response of these areas only to animal pictures, thus circumventing the multiple comparison problem, but potentially losing very interesting information about animal-selective areas somewhere else in the brain.

GLM (general linear model)-based approaches, which essentially aim to discover the most salient responses in the brain during a particular stimulation or task, are not the only way of analyzing fMRI data. There is also considerable interest in modelling the dependency of brain areas upon each other in analyses of functional or effective connectivity (the latter assessing also the directionality of this dependency). Furthermore, data-driven approaches, such as independent components analysis (ICA), have become very influential over the last decade, especially in the analysis of resting state data, which by definition do not lend themselves to the imposition of an experimental model.

2.4.4 Radionuclide imaging

The analysis of PET or SPECT studies of receptor occupancy is generally based on kinetic models of radiotracer distribution in the body (Laruelle et al. 2002). These models estimate the binding potential, that is, the ratio of the radioligand that is bound to the receptor to the free ligand. The basic assumption is that radioligand molecules will distribute across three relevant compartments: arterial blood, brain areas without relevant receptor concentrations (potential reference regions), and the target areas where the receptor density and/or occupation is to be measured. In order for this distribution to be reliably estimated, several measurements from the injection of the radiotracer are required, and classically this has involved arterial blood sampling to measure the kinetics of the radioligand in blood. However, such invasive models can be replaced by non-invasive models that do not require blood sampling if a suitable reference region can be identified. For example, the cerebellum is often used as a reference region for radiotracer studies of dopamine receptors in the striatum.

2.4.5 Diagnosis and prognosis through brain reading?

A recent approach is to use multi-voxel pattern analysis (MVPA) of imaging data for diagnostic purposes (Fig. 2.13). Most studies in this field have been based on structural imaging, using the volumes of different compartments of the brain. Examples are the distinction of patients with schizophrenia from healthy controls (Davatzikos et al. 2005), or adults with autism spectrum disorder (ASD) from controls

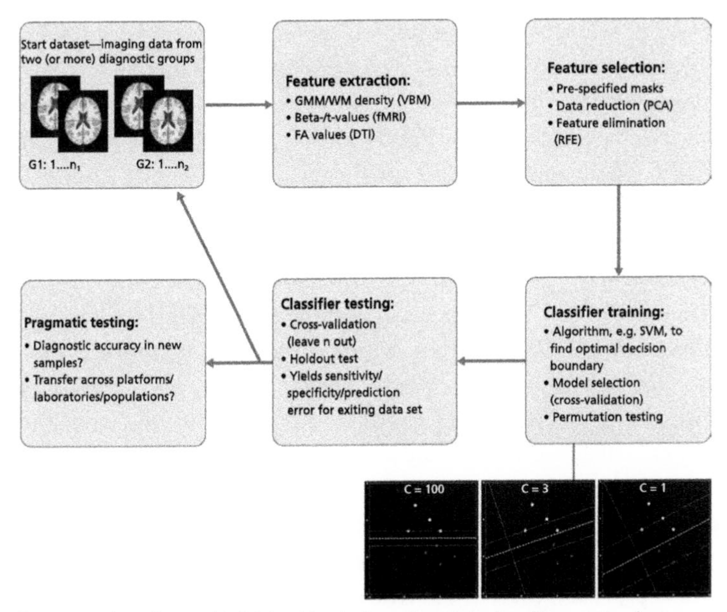

Fig. 2.13 Flowchart of MVPA of brain imaging data for diagnostic classification. Brain imaging data are obtained from two or more diagnostic groups and the relevant features extracted, as in standard univariate analysis. Because data reduction is crucial to the success of this procedure, various strategies for feature selection can be performed, including principal components analysis (PCA) and iterative elimination of non-discriminating features through recursive feature elimination (RFE) (De Martino et al. 2008). The preselected data are then fed into a classifier algorithm, which finds the optimal boundary between the two groups (e.g. the data points from the "bright" and "dark" groups in the right lower corner; figure panel courtesy of Professor Rainer Goebel). The approach is similar for higher numbers of features, except that hyperplanes rather than lines will constitute the decision boundaries. The performance of the trained classifier then has to be tested in independent data. Even a good performance on cross validation does not imply clinical relevance because the effects may be platform specific. An important pragmatic stage is therefore the testing of the classifier predictions in completely independent samples from other imaging centres, ideally, even with slightly altered measurement parameters to test robustness.

Reprinted from *Neuron*, 73,1, David E Linden, 'The Challenges and Promise of Neuroimaging in Psychiatry', pp. 8–22. Copyright (2012), with permission from Elsevier (Linden 2012).

(Ecker et al. 2010). In these studies, researchers were able to predict whether the scanned person had the disease or not with 80% to 90% accuracy, which is probably as high as the concordance between experienced clinicians. This classification is only as good as the underlying clinical diagnosis that is used to validate it. We will see in Chapter 7 that these very diagnostic categories that have been called into question because they are based on expert agreement between mental health clinicians rather than firm biological criteria or causal mechanisms, and that a new, dimensional approach based on the research domain criteria (RDOC) has been proposed, at least for research studies. Even if we can classify patients perfectly into the traditional diagnostic categories with brain imaging pattern analyses, these patterns may not tell us much about the underlying brain mechanisms.

However, real progress could be made if brain reading could allow psychiatrists to predict a mental disorder before the occurrence of clinical symptoms, especially if this might result in an early intervention to prevent the disease. Several studies have investigated this issue in groups of people who are known to be at high risk for developing a mental or neurodegenerative disorder, for example, prodromal schizophrenia (using structural imaging) (Koutsouleris 2009) or mild cognitive impairment (MCI) (using FDG-PET) (Mosconi et al. 2010). As yet, the practical relevance of brain reading for diagnostic or prognostic use has not been established. There are normally many more people without than with the disease in the population, and thus even a small proportion of false positive cases (unaffected individuals classified as affected by the disease by the brain reading procedure) will entail large numbers of falsely diagnosed individuals. Furthermore, brain reading procedures will only be of diagnostic use if they are stable cross scanners and laboratories (Klöppel et al. 2008).

References

Abhinav K, Yeh FC, Pathak S, et al. (2014) Advanced diffusion MRI fiber tracking in neurosurgical and neurodegenerative disorders and neuroanatomical studies: A review. *Biochim Biophys* Acta, **1842**(11):2286–97.

American Society of Neuroradiology (ASNR) (2014) *Practice parameter for the performance of computed tomography (CT) of the brain*. Available from: http://www.acr.org/~/media/ACR/Documents/PGTS/guidelines/CT_Brain.pdf. [8 February 2015].

Ashburner J, Friston KJ (2000) Voxel-based morphometry—the methods. *Neuroimage*, **11**(6 Pt 1):805–21.

Borogovac A, Asllani I (2012) Arterial Spin Labeling (ASL) fMRI: advantages, theoretical constrains, and experimental challenges in neurosciences. *Int J Biomed Imaging*, 2012:818456.

Dale AM, Fischl B, Sereno MI (1999) Cortical surface-based analysis. I. Segmentation and surface reconstruction. *Neuroimage*, **9**(2):179–94.

Davatzikos C, Shen D, Gur RC, et al. (2005) Whole-brain morphometric study of schizophrenia revealing a spatially complex set of focal abnormalities. *Arch Gen Psychiatry*, **62**(11):1218–27.

De Martino F, Valente G, Staeren N, Ashburner J, Goebel R, Formisano E (2008) Combining multivariate voxel selection and support vector machines for mapping and classification of fMRI spatial patterns. *Neuroimage*, **43**(1):44–58.

Deoni SC, Rutt BK, Arun T, Pierpaoli C, Jones DK (2008) Gleaning multicomponent T1 and T2 information from steady-state imaging data. *Magn Reson Med*, **60**(6):1372–87.

Detre JA, Wang J, Wang Z, Rao H (2009) Arterial spin-labeled perfusion MRI in basic and clinical neuroscience. *Curr Opin Neurol*, **22**(4):348–55.

Detre JA, Rao H, Wang DJ, Chen YF, Wang Z (2012) Applications of arterial spin labeled MRI in the brain. *J Magn Reson Imaging*, **35**(5):1026–37.

Ecker C, Marquand A, Mourão-Miranda J, et al. (2010) Describing the brain in autism in five dimensions—magnetic resonance imaging-assisted diagnosis of autism spectrum disorder using a multiparameter classification approach. *J Neurosc*, **30**(32):10612–23.

Friston KJ (2007) *Statistical parametric mapping : the analysis of functional brain images*. London: Academic.

Goebel R, Esposito F, Formisano E (2006) Analysis of functional image analysis contest (FIAC) data with brainvoyager QX: From single-subject to cortically aligned group general linear model analysis and self-organizing group independent component analysis. *Hum Brain Mapp*, **27**(5):392–401.

Homan P, Kindler J, Hauf M, Walther S, Hubl D, Dierks T (2013) Repeated measurements of cerebral blood flow in the left superior temporal gyrus reveal tonic hyperactivity in patients with auditory verbal hallucinations: a possible trait marker. *Front Hum Neurosci*, **7**:304.

Huettel SA, Song AW, McCarthy G (2014) *Functional magnetic resonance imaging*. third edn, Massachusetts, USA: Sinauer Associates, Inc., Publishers.

Jahng GH, Li KL, Ostergaard L, Calamante F (2014) Perfusion magnetic resonance imaging: a comprehensive update on principles and techniques. *Korean J Radiol*, **15**(5):554–77.

Jones DK (2011) *Diffusion MRI : theory, methods, and applications*. New York, Oxford: Oxford University Press.

Kamel N, Malik AS (2015) *EEG/ERP analysis : methods and applications*. Boca Raton: CRC Press, Taylor & Francis Group.

Klöppel S, Stonnington CM, Chu C, et al. (2008) Automatic classification of MR scans in Alzheimer's disease. *Brain*, **131**(Pt 3):681–9.

Koutsouleris N, Meisenzahl EM, Davatzikos C, et al. (2009) Use of neuroanatomical pattern classification to identify subjects in at-risk mental states of psychosis and predict disease transition. *Arch Gen Psychiatry*, **66**(7):700–12.

Laruelle M (2000) Imaging synaptic neurotransmission with in vivo binding competition techniques: a critical review. *J Cereb Blood Flow Metab*, **20**(3):423–51.

Laruelle M, Slifstein M, Huang Y (2002) Positron emission tomography: imaging and quantification of neurotransporter availability. *Methods*, **27**(3):287–99.

Leemans A, Jeurissen B, Sijbers J, Jones DK (2009) ExploreDTI: a graphical toolbox for processing, analyzing, and visualizing diffusion MR data. In 17th Annual Meeting of International Society for Magnetic Resonance in Imaging, Hawaii, p. 3537.

Linden DE (2012a) *The Biology of Psychological Disorders*. New York and Basingstoke: Palgrave Macmillan.

Linden DE (2012b) The challenges and promise of neuroimaging in psychiatry. *Neuron*, **73**(1):8–22.

Mills KL, Tamnes CK (2014) Methods and considerations for longitudinal structural brain imaging analysis across development. *Dev Cogn Neurosci*, **9**:172–90.

Morey RA, Petty CM, Xu Y, et al. (2009) A comparison of automated segmentation and manual tracing for quantifying hippocampal and amygdala volumes. *Neuroimage*, **45**(3):855–66.

Mori S and van Zijl PC (1995) Diffusion weighting by the trace of the diffusion tensor within a single scan. *Magn Reson Med*, **33**(1):41–52.

Mosconi L, Berti V, Glodzik L, Pupi A, De Santi S, de Leon MJ (2010) Pre-clinical detection of Alzheimer's disease using FDG-PET, with or without amyloid imaging. *J Alzheimers Dis*, **20**(3):843–54.

Neumann-Haefelin T, Wittsack HJ, Wenserski F, et al. (1999) Diffusion- and perfusion-weighted MRI. The DWI/PWI mismatch region in acute stroke. *Stroke*, **30**(8):1591–7.

Nikolaus S, Antke C, Müller H (2009) In vivo imaging of synaptic function in the central nervous system: I. Movement disorders and dementia. *Behav Brain Res*, **204**(1):1–66.

Nutt D and Nestor L (2013) *Addiction*. Oxford: Oxford University Press.

Petcharunpaisan S, Ramalho J, Castillo M (2010) Arterial spin labeling in neuroimaging. *World J Radiol*, **2**(10):384–98.

Poldrack RA, Mumford JA, Nichols TE (2011) *Handbook of functional MRI data analysis*. Cambridge: Cambridge University Press.

Ryken TC, Aygun N, Morris J, et al. (2014) The role of imaging in the management of progressive glioblastoma: a systematic review and evidence-based clinical practice guideline. *J Neurooncol*, **118**(3):435–60.

Seeman P, Madras B (2014) *Imaging of the human brain in health and disease.* Amsterdam and Boston: Academic Press/Elsevier.

Smith SM, Jenkinson M, Johansen-Berg H, et al. (2006) Tract-based spatial statistics: voxelwise analysis of multi-subject diffusion data. *Neuroimage*, **31**(4):1487–505.

Spencer AE, Uchida M, Kenworthy T, Keary CJ, Biederman J (2014) Glutamatergic dysregulation in pediatric psychiatric disorders: a systematic review of the magnetic resonance spectroscopy literature. *J Clin Psychiatry*, **75**(11):1226–41.

Supek S, Aine CJ (2014) *Magnetoencephalography: from signals to dynamic cortical networks*, Heidelberg: Springer.

Tzourio-Mzoyer N. Landeau B, Papthanassiou D, Crivello F, et al. (2002) Automated anatomical labeling of activations in SPM using a macroscopic anatomical parcellation of the MNI MRI single-subject brain. *Neuroimage*, **15**(1): 273–89.

Wijtenburg SA, Yang S, Fischer BA, Rowland LM (2015) In vivo assessment of neurotransmitters and modulators with magnetic resonance spectroscopy: Application to schizophrenia. *Neurosci Biobehav Rev*, **51**:276–95.

Wise RG, Harris AD, Stone AJ, Murphy K (2013) Measurement of OEF and absolute CMRO2: MRI-based methods using interleaved and combined hypercapnia and hyperoxia. *Neuroimage*, **83**:135–47.

Wong EC, Cox RW, Song AW (1995) Optimized isotropic diffusion weighting. *Magn Reson Med*, **34**(2):139–43.

Chapter 3

Techniques of neurophysiology and brain stimulation

Key points

- Electroencephalography (EEG) and magnetoencephalography (MEG) can measure synchronous potential changes in large numbers of neurons non-invasively
- Evoked potentials reflect the timing of neural activity with millisecond resolution
- Psychotropic drugs affect EEG and MEG signals
- Transcranial magnetic (TMS) and electrical stimulation (tDCS/tACS) modulate neural activity

3.1 EEG

Electroencephalography (EEG) measures cortical potential changes on the scalp and requires the spatial summation of large numbers of synchronous postsynaptic potentials for a sufficient signal-to-noise ratio. EEG uses electrodes (made of, for example, silver, lead, zinc, or platinum) and amplifiers, connected to a computer for data storage and analysis. The result is a visual picture of brain waves. Already, the inventor of the EEG, the German psychiatrist Hans Berger (1873–1941) observed that these brain waves change dramatically if the subject engages in mental activity, compared to rest. The resting rhythm is commonly in the alpha frequency (8–12 Hz), whereas cognitive activity and attention are accompanied by faster activity (beta: 12–30 Hz). Slow waves in the theta (3.5–7.5 Hz) and delta (<3.5 Hz) ranges occur during deep relaxation and sleep, but also during certain pathological states and as a consequence of psychotropic or narcotic drugs. Frequencies even higher than beta (gamma range: 30–100 Hz) have also been associated with cognitive activities although this high frequency activity has a relatively low amplitude and hence can only be extracted by computer analysis.

Many prescription and illicit drugs lead to changes in the EEG: for example, benzodiazepines increase beta activity; anticonvulsants can slow down the background rhythm from alpha to theta frequencies; and antipsychotics are also associated with the slowing of the EEG as well as increases in epileptiform activity. This activity resembles the synchronized sharp wave activity observed in patients with epilepsy, and sometimes even the characteristic spike/wave patterns (see Fig. 3.1). This epileptiform activity—which is most common (in about one third of treated patients) under clozapine and olanzapine—is thought to reflect a reduced seizure threshold. Antipsychotics and antidepressants do indeed carry a small risk of inducing seizures but in most patients these EEG changes will remain subclinical.

3.2 MEG

Another non-invasive method for measuring changes in synaptic activity is magnetoencephalography (MEG). MEG systems consist of arrays of sensors that pick up the magnetic field changes produced by the synaptic currents. Like EEG, MEG needs synaptic changes to occur synchronously in large numbers of neurons (at least in the order of 10,000s). However, the magnetic fields produced by synchronous changes in synaptic potentials are still very weak—10–100 Femtotesla (femto = 10^{-15}), about nine orders of magnitude below the earth's magnetic field—and extremely sensitive detectors, superconducting quantum interference devices (SQUIDs), are thus needed to record them. SQUIDs need to be bathed in liquid helium so that they are at their superconducting operating temperature of −269°C. The MEG laboratory needs to be protected against the influence of ambient electromagnetic noise as produced by railway lines or moving elevators.

Modern MEG systems have around 300 detectors. Because MEG has the advantage that the attenuation of magnetic signals depends only on the distance from the source, and not on the type of surrounding

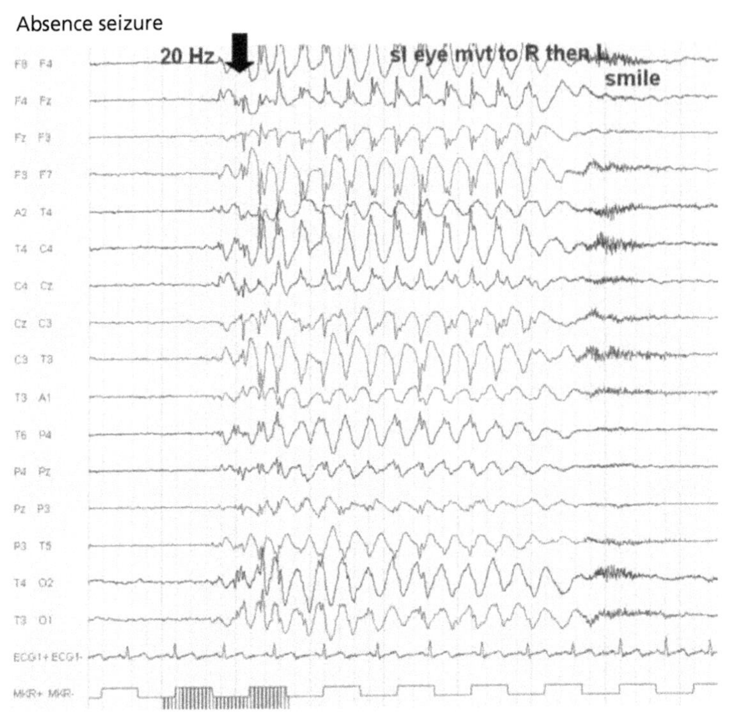

Fig. 3.1 Example of clinical EEG showing spike/wave complexes. This EEG was recorded during an absence seizure of an 11-year old female patient. During the seizure the patient became unresponsive and made slow lateral eye movements. The EEG shows a ca. 4 second episode of generalized spike/wave complexes (start indicated by arrow), preceded and followed by a normal trace. The bottom trace shows the calibration at 100 microvolts.
Reproduced from David Linden, *The Biology of Psychological Disorders*, 2011, with permission of Palgrave Macmillan.

tissue it allows for a more reliable reconstruction of cortical sources of scalp signals than EEG. MEG measurements are also less susceptible to contamination from muscle artefacts, such as those arising from the eyes, neck, and jaw than EEG. However, MEG only captures sources that are tangential to the scalp and thus does not necessarily reflect the same neural signals as EEG (see Fig. 3.2).

The main clinical application of MEG is in the evaluation of treatment-refractory epilepsy, where it can contribute to the localization of the epileptogenic focus. It can also be useful in presurgical

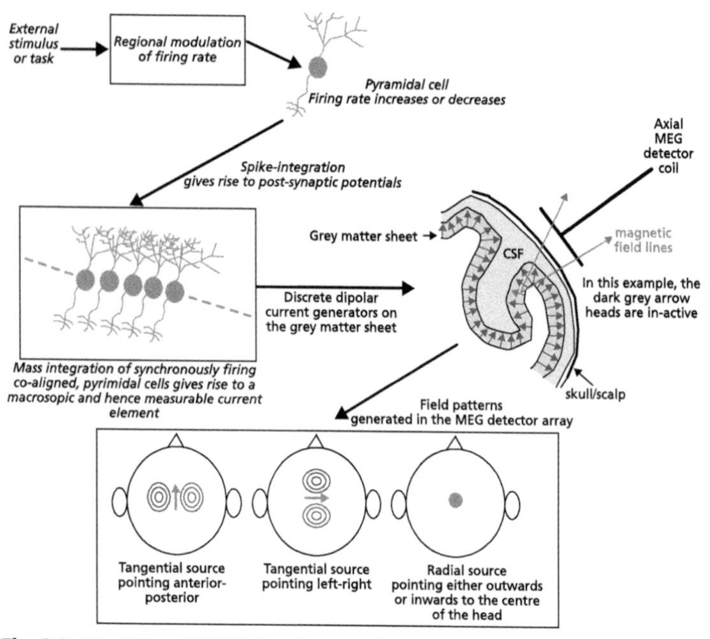

Fig. 3.2 Principles of MEG signal generation.

Senior, Carl, Tamara Russell, and Michael S. Gazzaniga, eds., *Methods in Mind*, figure 12.1, © 2006 Massachusetts Institute of Technology, by permission of The MIT Press.

functional mapping (in epilepsy and tumour surgery) in order to inform the surgeon about 'eloquent' brain areas in proximity to the operative target. For example, mapping of language networks with MEG has been proposed as an alternative to the Wada test (which involves transient suppression of one cerebral hemisphere by a barbiturate injection through the ipsilateral carotid artery) to ascertain lateralization of language functions. In psychiatry MEG is increasingly used as a research tool to investigate changes in functional brain networks.

3.3 Event-related/evoked potentials

EEG or MEG recordings during application of external stimuli or cognitive tasks allow for the computation of neural responses that are linked to the stimulus, task phases, or individual variations in behaviour. These are called evoked potentials (EP) in the case of responses to simple sensory stimuli and event-related potentials (ERP) for more complex, cognitive processes. The principle behind EP/ERP studies is that averaging over many trials will reduce the contribution of noise and preserve the signal that reflects neural processing. EPs have mainly been described for the visual, auditory, and tactile domain, but can be measured for all sensory channels. EPs are normally described as positive or negative going (denoted with the letters P or N), according to the direction of the wave in standard referencing procedures. The letter is followed by a number that denotes the latency in milliseconds or the position in a sequence of positive or negative deflections.

A classical application of ERPs uses so-called oddball paradigms, where a train of regular 'standard' stimuli (e.g. tones of a particular frequency) is disrupted by a deviant 'oddball' (a tone at a different frequency). These oddballs are associated with the P3 or P300 response, which can be elicited in all sensory modalities, and the auditory 'mismatch negativity'. The P300 is probably the most widely studied neurophysiological biomarker of mental disorders. The mismatch negativity (MMN) is also of great neuropsychiatric interest, for example, because of its modulation by glutamate antagonists (Stephan et al. 2006).

One limitation of all measures derived from EEG or MEG is that they only capture the correlates of brain activity recorded on the scalp. This limitation can be overcome to a certain extent by the use of algorithms that try and find the most likely location of the sources of this activity in the brain. However, this approach requires disentangling the unknown number of sources of an unknown number of neural processes whose signals overlap in scalp recordings. The combination of EEG or MEG with other imaging modalities—mainly structural and functional MRI—is therefore crucial in order to improve the fidelity of source localizations of electromagnetic activity. Special EEG electrode caps and amplifiers are available for simultaneous recording with MRI/fMRI. This approach, which combines the high spatial resolution of MRI (millimetres) with the high temporal resolution of EEG (milliseconds), achieves the optimal non-invasive spatio-temporal mapping of brain function, but also poses considerable challenges in terms of data post-processing and artefact correction (Mulert and Lemieux 2010).

3.4 Transcranial stimulation

Transcranial magnetic stimulation (TMS) induces electrical fields in the brain non-invasively. TMS works according to the laws of electromagnetic induction. The TMS apparatus is a bank of capacitors that discharge a strong (up to 10,000 Amperes (A)) and very brief current (approx. 200 microseconds) into a coil that is held over the head. This current generates a magnetic field that induces an electric field in the tissue under the coil. If this tissue is conductive (as is the case for nervous tissue) this electric field will lead to an electric current, which can cause local membrane de- or hyperpolarization. Depending on the orientation of the coil and the stimulated neurons, and of the stimulation parameters, the effects of TMS can be excitatory or inhibitory. For example, single pulse stimulation over the hand area of the primary motor cortex normally leads to a movement in the contralateral hand, an excitatory effect. Similarly, stimulation over the visual cortex may induce transient visual impressions, called phosphenes. Conversely, repetitive stimulations at low frequencies (LF) (e.g. 1 Hz) or the more recently developed 'theta burst' (three cycles per second of fast bursts of five pulses) have inhibitory effects. TMS is used in clinical neurophysiology to test the integrity of corticospinal motor pathways and assess plasticity in the motor system. Inhibitory TMS protocols can also be used experimentally to induce functional lesions. TMS is thus a unique method for the systematic assessment of the functions of specific brain areas (Sack and Linden 2003).

In psychiatry, repetitive TMS (rTMS) is also used as experimental treatment method for hallucinations and depression. Putatively inhibitory (LF) or excitatory (high frequency (HF)) (approx. 10Hz) frequencies are used, depending on whether the targeted area is supposed to be hyper- or hypoactive. For depression, HF-TMS over the left frontal lobe and LF-TMS over the right frontal lobe are the most commonly used protocols, and recent meta-analyses support the efficacy of this method as an add-on treatment for depression, at least for short-term effects (see Chapter 10) (Berlim et al. 2013).

Electric fields in the brain can also be induced by direct electrical stimulation through electrodes placed on the scalp. Brief high intensity stimulation—inducing electric fields in the brain with a field strength of around 200V/m—is used for electroconvulsive treatment (ECT) (see Chapter 5). Stimulation with low currents is used in transcranial direct current stimulation (tDCS) or transcranial alternating current stimulation (tACS) and induces weak electric fields in the brain (amplitudes of about 0.2–2 V/m). TDCS is postulated to increase the excitability of the area under the anode and decrease that of the area under the cathode. However, its effects are much weaker than those of TMS and thus no overt motor stimulation is achieved under the anode (the electric field corresponding to the motor threshold is in the range of 30 V/m to 130 V/m). However, tACS over visual cortex can induce phosphenes, and various effects of cognitive enhancement have been described in preliminary studies of tDCS, and evaluation of tDCS protocols for psychiatric treatment, especially for depression, is underway (see Chapter 10).

References

Berlim MT, van den Eynde F, Daskalakis ZJ (2013) Clinically meaningful efficacy and acceptability of low-frequency repetitive transcranial magnetic stimulation (rTMS) for treating primary major depression: a meta-analysis of randomized, double-blind and sham-controlled trials. *Neuropsychopharmacology*, **38**(4):543–51.

Mulert C and Lemieux L (eds) (2010) *EEG - fMRI: Physiological Basis, Technique, and Applications*, Heidelberg: Springer.

Sack AT and Linden DE (2003) Combining transcranial magnetic stimulation and functional imaging in cognitive brain research: possibilities and limitations. *Brain Res Brain Res Rev*, **43**(1):41–56.

Stephan KE, Baldeweg T, Friston KJ (2006) Synaptic plasticity and dysconnection in schizophrenia. *Biol Psychiatry*, **59**(10):929–39.

Chapter 4

Clinical indications of neuroimaging in psychiatry

Key points

- Neuroimaging can help rule out defined organic pathologies and differentiate between organic causes of mental illness
- MRI is the preferred modality for the diagnostic workup of dementia
- SPECT and PET can be helpful for the differentiation between types of dementia
- Current guidelines do not recommend routine scanning in patients with psychosis, learning disabilities, or conduct disorder
- However, clinicians need to be aware that 'classical' psychiatric syndromes can also be the consequence of a brain lesion
- Mental and behavioural disorders resulting from specific brain disorders, for example, epilepsy, are the remit of neuropsychiatry

4.1 Dementia and delirium

4.1.1 Dementia

Dementia is defined in the *The ICD-10 Classification of Mental and Behavioural Disorders* (ICD) as a 'syndrome due to disease of the brain, usually of a chronic or progressive nature, in which there is disturbance of multiple higher cortical functions, including memory, thinking, orientation, comprehension, calculation, learning capacity, language, and judgement' (http://apps.who.int/classifications/icd10/browse/2010/en#/F00, Accessed on 30 May 2015). Dementia is distinguished from a primary learning disability by the decline from a previously attained level of cognitive functioning. Although dementia is a purely clinical diagnosis, establishing the underlying brain disorder requires a range of diagnostic tests, including neuroimaging. In the UK, the current guidelines for the National Health Service, produced by the National Institute for Health and Care Excellence (NICE), recommend the use of 'structural imaging to exclude other cerebral pathologies and help establish the subtype', although it 'may not always be needed in those presenting with moderate to severe dementia, if the diagnosis is already clear.' The preferred imaging modality is MRI because it is more sensitive than CT to the brain changes associated with the commonest causes of dementia (Alzheimer's and vascular dementia) and also much more helpful than CT in establishing the rarer underlying diagnoses, for example, fronto-temporal dementia or inflammatory diseases. For the differentiation between Alzheimer's disease (AD), vascular dementia (VD), and frontotemporal dementia (FTD) it can also be useful to identify the characteristic regional deficits in perfusion and glucose utilization with hexamethylpropyleneamine oxime (HMPAO)-SPECT or FDG-PET (NICE 2014) (see Fig. 4.1).

The status of biomarkers (which are mainly derived from neuroimaging and neurochemistry) has recently been strengthened by the revision of the diagnostic criteria for AD. Although the inclusion of biomarker tests is still not recommended for routine clinical use

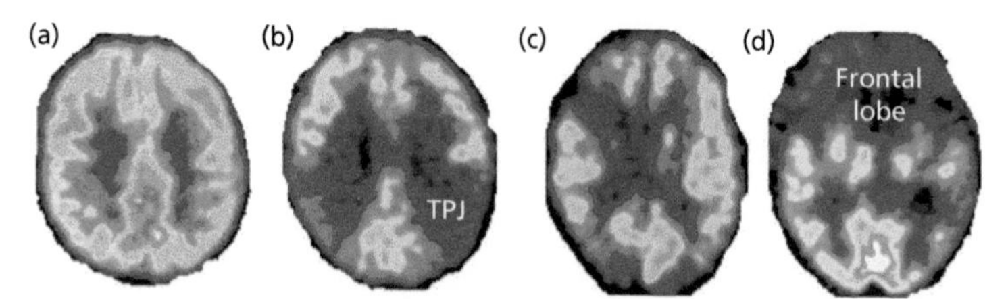

Fig. 4.1 PET distinguishes different types of dementia. Maps of glucose metabolism, measured with FDG-PET, in (a) a healthy control (aged 55); (b) a patient with Alzheimer's Disease (AD) (aged 60); (c) a patient with vascular dementia (VD) (aged 50); and (d) a patient with frontotemporal dementia (FTD) (aged 69). Note that hypometabolism is most marked in the temporoparietal junction (TPJ) in AD and in the frontal lobe in FTD, but more diffuse in VD.

Reproduced from David Linden, *The Biology of Psychological Disorders*, 2011, with permission of Palgrave Macmillan.

it is recognized that a diagnosis of AD will receive added certainty through biomarker evidence of amyloid deposition and neurodegeneration (McKhann et al. 2011). The former can be provided through measuring amyloid-beta (Aβ) levels in the CSF (which are decreased) or through demonstration of increased levels of amyloid binding on PET. The evidence for neurodegeneration would be provided through increased tau protein levels in CSF, cerebral hypometabolism on FDG-PET, or atrophy on MRI (Cohen and Klunk 2014).

PET imaging of amyloid burden using the ^{11}C-based tracer Pittsburgh Compound B (PiB) or one of the recently introduced ^{18}F-based tracers (Florbetaben, Florbetapir, Flutemetamol) has a good sensitivity for AD compared to other causes of dementia but is not very specific—it shows amyloid burden also in a considerable portion of patients with mild cognitive impairment, and even in many healthy elderly people (and in people with Down syndrome). However, it can differentiate AD- from FTD-pathology even better than FDG-PET (Cohen and Klunk 2014).

If dementia with Lewy bodies (DLB) is suspected, the associated dopaminergic deficit can be documented with a DaTSCAN (TM) (see also Chapter 2). As in Parkinson's disease, nigrostriatal dopaminergic neurons degenerate in DLB, which results in less binding capacity for the radiolabelled DAT ligand ^{123}I-2-carbomethoxy-3-(4-iodophenyl)-N-(3-fluoropropyl) nortropane (FP-CIT). A lower signal on the DaTSCAN would therefore be indicative of DLB. The main indication for the DaTSCAN, however, is still the differentiation between PD (dopaminergic deficit) and other movement disorders.

EEG is not part of the routine workup of dementia. It can be useful to support a diagnosis of Creutzfeldt Jakob Disease (CJD), especially if patients show the typical periodic sharp wave complexes. Patients with AD and other forms of dementia also have an increased risk of developing epilepsy (Friedman et al. 2012), and thus, EEG will often be required to help investigate the causes of seizures and inform their management.

4.1.2 **Delirium**

Delirium is a syndrome of acute confusion and disturbed cognitive functioning and behaviour that generally arises as part of a physical

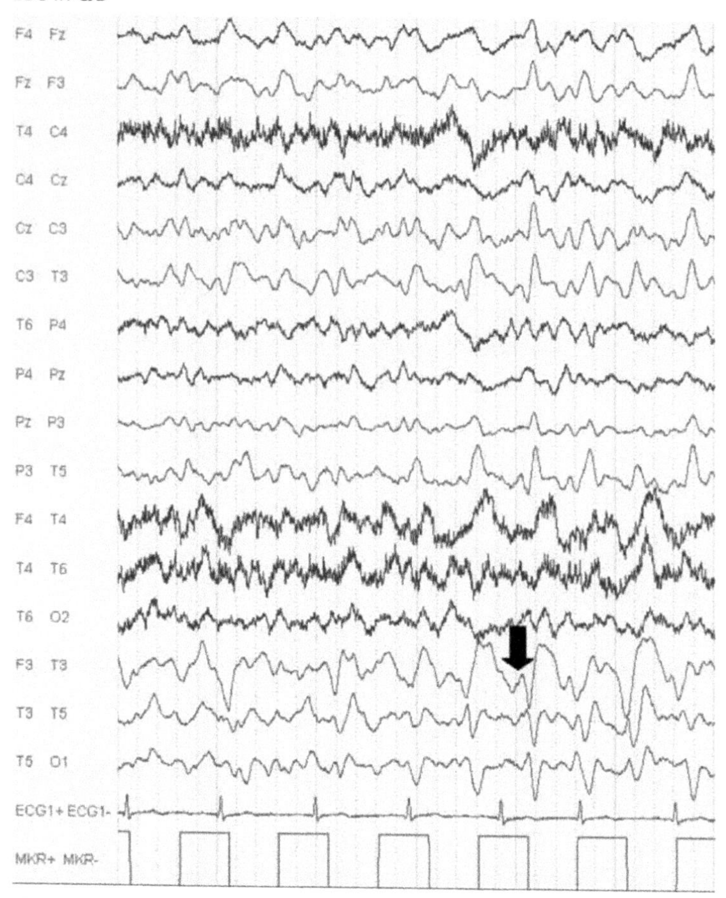

Fig. 4.2 EEG in Creutzfeldt Jakob Disease). The EEG of this 55-year-old patient with rapidly progressive cognitive decline who was eventually diagnosed with CJD shows generalized slowing and left lateralized (odd numbers denote the left side of the head, even numbers the right) biphasic and triphasic (arrow) waves. Periodic bi- or triphasic waves are typical of CJD, although they are also often observed in other encephalopathies, most notably those of hepatic and other metabolic origin. Vertical dotted lines denote intervals of 200 ms and the bottom trace shows the amplitude calibration at 100 Microvolts.

Reproduced from David Linden, *The Biology of Psychological Disorders*, 2011, with permission of Palgrave Macmillan.

illness, as a result of intoxication or during withdrawal from drugs or alcohol. Because it can indicate a medical emergency, an immediate and comprehensive physical workup is needed. However, imaging is not routinely part of this. Clinical guidelines (Young and George 2006) suggest that CT scanning should be performed in patients in whom an intracranial lesion is suspected because of focal neurological signs, a head injury, a fall, or clinical evidence of raised intracranial pressure (for example, vomiting or papilloedema). Similarly, EEG is not part of the routine workup but can be useful in differentiating delirium from non-convulsive status epilepticus or temporal lobe epilepsy (which may be accompanied by characteristic epileptic discharges).

4.2 **Patients with a first manifestation of psychosis or affective symptoms**

Protocols for the diagnostic assessment of patients with a first manifestation of psychiatric illness are widely debated. On the one hand, psychosis or affective symptoms can be early indicators of an underlying brain lesion. One the other hand, large scale screening studies of psychiatric patients may not be cost effective and bring up incidental findings that can cause undue concerns. Before we review the epidemiological evidence for clinically relevant imaging findings in psychiatric patient populations and discuss the current clinical guidelines it is worth reviewing some case studies of the possible relationship between brain lesions and psychiatric symptoms.

Case 1

'This 56-year-old right-handed homemaker had progressive apathy, social withdrawal, and poor self-care for three years and was admitted to a psychiatric facility for depression. Because she was unresponsive to appropriate antidepressant medications, a CT scan was taken of the head. This study revealed an enhancing, 8-cm, medial bifrontal mass' (see Fig. 4.3). 'Total excision of a benign transitional-type meningioma led to rapid and dramatic improvement, and four months after the operation she was animated, cheerful, and motivated to resume her previous life' (Filley and Kleinschmidt-DeMasters 1995).

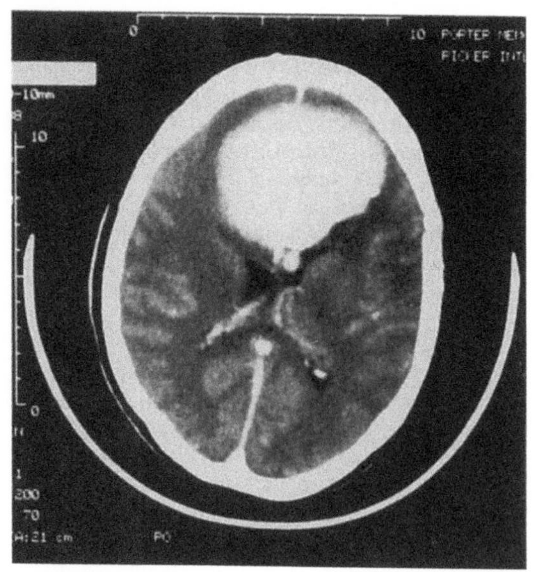

Fig. 4.3 CT scan of Case 1.

Reproduced from Filley and Kleinschmidt-DeMasters 1995 with permission from BMJ Publishing Group Ltd.

Case 2

An 18-year-old white female was referred by her school because she appeared to be at high risk of psychosis. According to her family, two years previously she started to withdraw from social activities and resented participating in work groups and talking in public. One year later, she felt strangers were staring and laughing at her for no reason, and that the world around her had changed. At her initial evaluation she felt that people were intimidating her and sometimes imagined that television programmes were sending special messages to her, although she was not certain about this and could not describe these messages. This patient was neurologically normal and had an above average IQ. She was initially diagnosed with a prodromal syndrome of schizophrenia, but symptoms became rapidly more severe. A routine MRI scan revealed a tumour in the left temporal lobe (see Fig. 4.4), which was surgically removed. Histopathologically it was a dysembryoplastic neuroepithelial tumor (DNET), a generally benign glial-neural neoplasm. The patient's psychotic symptoms subsequently improved

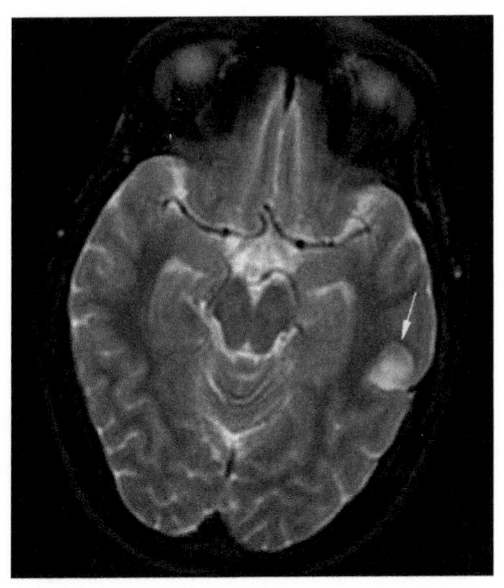

Fig. 4.4 T2-weighted MR scan of Case 2.

Reprinted from *Schizophrenia Research*, 144, 1–3, 'Brain tumor in a patient with attenuated psychosis syndrome', pp. 151–2. Copyright (2013), with permission from Elsevier.

under treatment with risperidone and cognitive behavioural therapy (CBT), but she remained socially withdrawn (Zugman et al. 2013).

Case 3

A 48-year-old female with a history of major depressive disorder suffered a relapse of depressive symptoms, which was not responsive to her usual combination of paroxetine and risperidone. She also developed amenorrhoea, and laboratory testing revealed hyperprolactinemia and low oestrogen. A pituitary prolactinoma was found on MRI. Both the affective symptoms and the menstrual problems improved after treatment with dostinex, a dopamine agonist, and aripiprazole, a partial dopamine agonist (Liao and Bai 2014).

These cases illustrate the possibility that classical psychiatric clinical presentations, such as depression or psychosis, arise from brain tumours. Cases 1 and 3 had additional clinical characteristics (non-response to treatment in Cases 1 and 3, and amenorrhoea in Case 3) that may have suggested an atypical cause. These cases underline the importance of detailed history taking and physical examination for the diagnostic workup of patients presenting primarily with psychiatric symptoms. Case 2 was only detected through routine screening. This begs the question whether clinically significant imaging findings are more common in patients with psychotic disorders than in the general population. This question was investigated by Sommer et al. (2013). They compared the MRI scans of 656 patients with psychosis and 722 healthy controls. Both groups had around 10% abnormalities that were classified as clinically relevant, but none was deemed to be a substrate for an organic psychosis, and brain tumours were only observed in 0.2% of the patients and in 0.3% of the controls. The authors concluded that neuroimaging would not be a cost-effective component of the routine workup of psychotic disorders. This assessment is in keeping with the most recent UK recommendations: 'Structural neuroimaging, using methods called magnetic resonance imaging (MRI) or computed axial tomography (CT) scanning, is not recommended for use routinely to examine all people who have had a first episode of psychosis' (NICE 2008).

Molecular imaging with PET or SPECT has presently no place in the diagnostic workup of psychosis or affective disorders. Assessing the dopamine system with DAT ligands can have some clinical use in distinguishing Parkinsonism caused by antipsychotic drugs from PD (Brigo et al. 2014) because the former is caused by a post-synaptic dopaminergic blockade, whereas the latter is caused by dopamine depletion. Abnormalities on DAT imaging would therefore only be expected in PD, but not in antipsychotic-induced Parkinsonism. However, some patients with drug-induced Parkinsonism also have abnormal DAT imaging findings (and some patients with neuropathologically validated PD show no abnormalities) and the diagnostic utility of this molecular imaging technique for the distinction between different Parkinsonian syndromes therefore still requires further studies.

4.3 Behavioural problems and learning disability

4.3.1 Learning disability

Learning disability or intellectual disability per definition affects approx. 2% of the population if the cutoff point is set at IQ = 70 (see Chapter 6). Causes include birth defects (for example, perinatal hypoxia) and genetic syndromes, but in many cases the aetiology remains unclear. Comorbidity with other physical problems and neurological conditions, particularly epilepsy, is common. In patients who present with learning disability and epilepsy neuroimaging is generally recommended. MRI is the method of choice unless sedation, which is commonly needed for MRI in young children, can be avoided by performing CT (NICE 2012).

In patients with LD without comorbid neurological disorder, the indication for imaging is less clear. Although about 30% of patients with LD have abnormal imaging findings, most of these probably are not causal and MRI only helps in establishing a diagnosis in 0% to 4% of cases. Thus, MRI is not necessarily required for the routine clinical workup of LD, but should be considered for patients with focal neurological signs or microcephaly (Moeschler and Shevell 2006).

4.3.2 Behavioural problems

Up to 10% of children have behavioural problems that warrant clinical evaluation and may fulfil the diagnostic criteria for conduct disorder, which consists in the persistent violation of social norms and basic rules of behaviour, or attention deficit/hyperactivity disorder, which is characterized by the symptom clusters of inattention and impulsivity. Although unusually aggressive and antisocial behaviour can, in very rare cases, be caused by brain tumours and be the main clinical manifestation of associated seizure activity (Nakaji et al. 2003) brain imaging is not part of the standard clinical workup of conduct problems. However, clinicians should keep the possibility of an organic cause in mind and consider imaging especially in clinically unusual cases, for example, those with rapid onset or severe fluctuations of behavioural problems, and certainly in any patient with comorbid seizures. Sudden onset of aggressive and antisocial behaviour in adults can also indicate an underlying brain lesion, for example, a tumour in the orbitofrontal cortex, or a frontal lesion as in the famous case of Phineas Gage (Haas 2001). In order to rule out such an 'acquired sociopathy' (Gao et al. 2009) with underlying organic lesion imaging should be considered in cases with recent and otherwise unexplained behavioural problems.

4.4 Neuropsychiatric patients

Patients who come to the attention of psychiatrists (and sometimes dedicated neuropsychiatrists) because of mental health problems arising from a neurological disease will often already have undergone some type of imaging to establish the neurological diagnosis and localization of a lesion, for example, in cases of stroke, brain tumour, traumatic brain injury, or multiple sclerosis (MS). The indication for renewed imaging is commonly driven by a deterioration in the neurological status, which can, for example, indicate an exacerbation of MS or a suspected recurrence of a brain tumour, but a deterioration in the mental status may also index an observable change in the brain and thus warrant a new scan. For example, relapses of MS are often associated with depression and anxiety (Moore et al. 2012), and a first episode of depression in a patient with established MS can indicate a flare-up of the inflammatory process.

However for some neurological diagnoses that are commonly associated with psychiatric morbidity neuroimaging is not part of the standard workup. For example, the diagnosis of Parkinson's disease (PD) can generally be based on clinical criteria (NICE 2006 and Wippold et al. 2015). In the workup of Huntington's disease, the suspected diagnosis is established on the basis of the clinical assessment and family history and confirmed by genetic testing, and again many patients will not routinely get brain scans. If patients with these movement disorders deteriorate in their mental state brain imaging may be needed to chart the progression of illness or, in suspected PD, revisit the diagnosis. A classical example would be a patient with Parkinsonian motor symptoms who also develops dementia early in the course of the illness, which may warrant further tests to rule out DLB. Although clinical imaging markers for DLB have not yet been established, PET imaging of amyloid load is a promising tool for the differentiation of DLB from PD because the cortical amyloid burden seems to be high in DLB but low in PD (Siderowf et al. 2014).

References

Brigo F, Matinella A, Erro R, Tinazzi M. (2014) [^{123}I]FP-CIT SPECT (DaTSCAN) may be a useful tool to differentiate between Parkinson's disease and vascular or drug-induced parkinsonisms: a meta-analysis. *Eur J Neurol*, **21**(11):1369–e90.

Cohen AD, Klunk WE (2014) Early detection of Alzheimer's disease using PiB and FDG PET. *Neurobiol Dis*, **72**, Pt. A:117–22.

Filley CM, Kleinschmidt-DeMasters BK (1995) Neurobehavioral presentations of brain neoplasms. *West J Med*, **163**(1):19–25.

Friedman D, Honig LS, Scarmeas N (2012) Seizures and epilepsy in Alzheimer's disease. *CNS Neurosci Ther*, **18**(4):285–94.

Gao Y, Glenn A, Schug R, Yang Y, Raine A (2009) The neurobiology of psychopathy: a neurodevelopmental perspective. *Can J Psychiatry*, **54**(12):813–23.

Haas LF (2001) Phineas Gage and the science of brain localisation. *J Neurol Neurosurg Psychiatry*, **71**(6):761.

Liao WT, Bai YM (2014) Major depressive disorder induced by prolactinoma—a case report. *Gen Hosp Psychiatry*, **36**(1):125.e121–2.

McKhann GM, Knopman DS, Chertkow H, et al. (2011) The diagnosis of dementia due to Alzheimer's disease: recommendations from the National Institute on Aging-Alzheimer's Association workgroups on diagnostic guidelines for Alzheimer's disease. *Alzheimers* Dement,7(3):263–9.

Moeschler JB, Shevell M (2006) Clinical genetic evaluation of the child with mental retardation or developmental delays. *Pediatrics*,**117**(6):2304–16.

Moore P, Hirst C, Harding KE, Clarkson H, Pickersgill TP, Robertson NP (2012) Multiple sclerosis relapses and depression. *J Psychosom Res*, 73(4):272–6.

National Institute for Health and Care Excellence (NICE) (2015) *Dementia diagnosis and assessment*, NICE pathways. Available from: http://pathways.nice.org.uk/pathways/dementia#path = view%3A/pathways/dementia/dementia-diagnosis-and-assessment.xml&content = view-index [8 August 2015].

National Institute for Health and Care Excellence (NICE) (2006) *Parkinson's disease: Diagnosis and management in primary and secondary care*, NICE clinical guideline 35. Available from: http://www.nice.org.uk/guidance/cg35/resources/guidance-parkinsons-disease-pdf [8 September 2015].

National Institute for Health and Care Excellence (NICE) (2008) *Structural neuroimaging in first-episode psychosis*, NICE technology appraisals guidance (TA136). Available from: https://www.nice.org.uk/guidance/ta136 [10 September 2015].

National Institute for Health and Care Excellence (NICE) (2012) *The epilepsies: the diagnosis and management of the epilepsies in adults and children in primary and secondary care*, NICE clinical guideline 137. Available from: https://www.nice.org.uk/guidance/cg137 [8 September 2015].

Nakaji P, Meltzer HS, Singel SA, Alksne JF (2003) Improvement of aggressive and antisocial behavior after resection of temporal lobe tumors. *Pediatrics*, **112**(5):e430.

Siderowf A, Pontecorvo MJ, Shill HA, et al. (2014) PET imaging of amyloid with Florbetapir F 18 and PET imaging of dopamine degeneration with 18F-AV-133 (florbenazine) in patients with Alzheimer's disease and Lewy body disorders. *BMC Neurol*, **14**:79.

Sommer IE, de Kort GA, Meijering AL, et al. (2013) How frequent are radiological abnormalities in patients with psychosis? A review of 1379 MRI scans. *Schizophr Bull*,**39**(4):815–19.

Wippold FJ, Brown DC, Broderick DF, et al. (2015) ACR Appropriateness Criteria Dementia and Movement Disorders. *J Am Coll Radiol*, **12**(1):19–28.

Young L, George J (1997) *Guidelines for the diagnosis and management of delirium in the elderly*, British Geriatrics Society. Available from: https://www.rcplondon.ac.uk/sites/default/files/concise-delirium-2006.pdf [8 September 2015].

Zugman A, Pan PM, Gadelha A, et al. (2013) Brain tumor in a patient with attenuated psychosis syndrome. *Schizophr Res*, **144**(1–3):151–2.

Chapter 5

Clinical indications of EEG in psychiatry

Key points

- EEG is needed when an epileptic origin of an abnormal mental state is suspected
- EEG can help with the differentiation between epileptic and non-epileptic attacks and the diagnosis of other functional neurological disorders
- EEG is a central part of the diagnosis of sleep disorders
- EEG is needed to monitor the seizures induced by electroconvulsive treatment (ECT)

5.1 Differential diagnosis of altered states of consciousness

5.1.1 Ictal events

Identifying whether a period of abnormal behaviour or a state of altered consciousness is produced by underlying epileptic activity or whether it arises as part of a mental (psychotic or dissociative) disorder can be a diagnostic challenge. Detecting characteristic epileptic discharges during such an attack ('ictal') would lend strong support to an epileptic origin and inform therapeutic decisions (in favour of anticonvulsant treatment). Such neurophysiological support for an epileptic origin of abnormal behaviour can also have considerable forensic importance because epilepsy is increasingly being cited as a reason for decreased culpability in criminal cases. However, contrary to some historic beliefs (Linden 2014), there is little evidence for a causal link between epilepsy and offending behaviour (Fazel et al. 2011), and ictal violence is rare. To determine whether violent behaviour was the result of an epileptic seizure, documentation of epileptic automatisms with aggressive acts during video-EEG is required, in addition to a careful clinical workup of the patient's seizure patterns (Treiman 1986).

The epilepsies also have important psychiatric comorbidities, which can occur during (ictal) or around the time of seizures (periictal) or in seizure-free intervals (interictal). Amongst the classical ictal phenomena, complex partial status epilepticus is sometimes associated with hallucinations or other psychotic phenomena (Elliott et al. 2009). A significant proportion of people admitted to casualty departments with confusion have EEG abnormalities, and non-convulsive status epilepticus is one of the diagnostic possibilities to consider (Zehtabchi et al. 2013). EEG (ideally video-EEG) recordings can be helpful in identifying the epileptic origin of altered mental states (see Fig. 5.1). Whereas the treatment of ictal psychosis would consist in the optimization of seizure control and possibly short-term sedation, the treatment of psychosis of non-epileptic origin would generally involve antipsychotic medication. Another neuropsychological syndrome for which EEG may aid diagnosis is transient amnesia, which can have an epileptic origin and respond to anticonvulsant medication (Asadi-Pooya 2014).

5.1.2 Periictal (pre-/postictal) events

Examples of periictal neuropsychiatric symptoms include premonitory depression and irritability preceding seizures and transient postictal psychotic states. Some of these periictal phenomena seem to result from subclinical seizure activity. For example, for postictal psychosis, which can resemble transient states of paranoid or affective psychosis of other origins, invasive EEG recordings have demonstrated ongoing limbic seizure activity. Thus they may be regarded as a form of non-convulsive status epilepticus rather than a truly post-ictal phenomenon, although in most cases the scalp EEG will not show any epileptiform discharges.

5.1.3 Interictal events

EEG can also be helpful in the workup of interictal psychotic symptoms. Up to 20% of epilepsy patients can be affected by the interictal schizophrenia-like psychosis of epilepsy (SLPE), which is

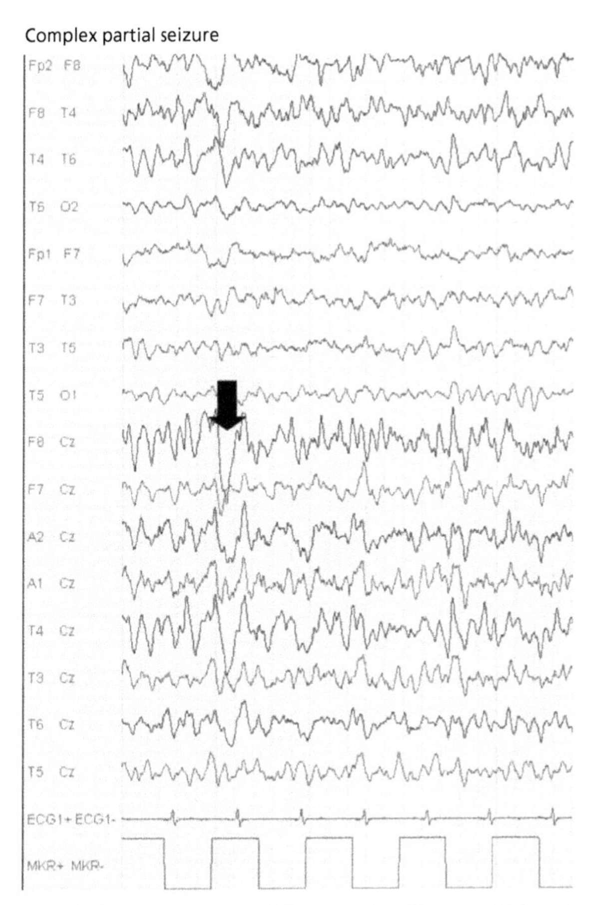

Fig. 5.1 EEG during a complex partial seizure in a 29-year-old female patient with a history of Rasmussen encephalitis, a rare autoimmune brain inflammation. She developed complex partial seizures during which she is unresponsive and turns her head backwards. On this occasion the clinical event was preceded by right fronto-temporal sharp waves (even electrode numbers denote the right side of the head, odd numbers the left).

Reproduced from David Linden, *The Biology of Psychological Disorders*, 2011, with permission of Palgrave Macmillan.

characterized by delusions and visual hallucinations. It arises relatively late in the course of the illness (generally over ten years) and individual episodes can last for several years. Although traditionally psychiatrists had considered epilepsy to be a protective factor against schizophrenia—a view, which motivated Ladislaus von Meduna to develop ECT (Shorter and Healy 2007)—there is now increasing evidence that subclinical epileptic activity can contribute even to the interictal psychoses, although it generally will not be picked up by conventional scalp EEG. A normal EEG therefore does not rule out SLPE, but more detailed investigations such as sleep EEG may help the diagnostic process in a patient with psychotic symptoms and clinical suspicion of epilepsy. Because the pathophysiology of SLPE is unclear treatment is symptom-focused and generally involves optimization of seizure control, antipsychotic medication, and psychosocial support (Adachi et al. 2013).

5.1.4 Non-epileptic attacks

EEG is an important diagnostic tool for the differentiation between epileptic and non-epileptic seizures. Non-epileptic (also termed 'psychogenic or 'functional') seizures are 'paroxysmal behavioural, motor, sensory, autonomic, cognitive, and/or emotional changes that resemble or are mistaken for epileptic seizures' (Mostacci et al. 2011). The most famous descriptions go back to the nineteenth century French neurologist Jean-Martin Charcot, but non-epileptic seizures are still common, contributing up to 40% of referrals to tertiary epilepsy centres for evaluation of treatment-refractory seizures. Although in many cases reasonable confidence as to the epileptic or non-epileptic origin of these attacks can be derived from a clinical analysis, based on careful history taking, observations, and witness accounts of seizures, there will be cases where the diagnostic uncertainty can only be resolved through video-EEG recording. A normal ictal video-EEG and observation of clinical features of non-epileptic seizures are the criteria for a diagnosis of documented psychogenic non-epileptic seizures (PNES) (LaFrance et al. 2013).

A particular diagnostic challenge is the assessment of patients with clinically suspected non-epileptic seizures that arise from behavioural sleep. (This term is used to denote periods during which a person appears to be asleep; it is further classified as 'pseudosleep' if the person has a waking EEG.) Clinical experience suggests that non-epileptic seizures do not arise from biological sleep, and scalp-EEG can clarify whether the period of behavioural sleep preceding a seizure was accompanied by normal waking EEG activity (see Section 5.3), corroborating the clinical suspicion of a non-epileptic seizure origin. This diagnostic distinction between epileptic and non-epileptic seizures has huge therapeutic impact because anticonvulsants are ineffective or even harmful in non-epileptic seizure disorder, which is classified as a dissociative (conversion) disorder in the ICD-10 (F44.5) (World Health Organization 1992), requires assessment of any underlying psychiatric morbidity and can be amenable to psychological intervention.

A similar diagnostic problem arises in the distinction between syncope (which is associated with marked reductions in cerebral perfusion) and psychogenic pseudosyncope (which is not). Pseudosyncope is an apparent transient loss of consciousness without organic correlate that is distinguished from functional seizures by the absence of motor behaviour. Although EEG does not measure blood flow directly, it can detect the neurophysiological consequences of hypoperfusion, and during syncope generally shows slowing, including marked delta waves. Conversely, EEG during psychogenic pseudosyncope would show a normal waking alpha rhythm (Raj et al. 2014). The provocation methods commonly used in the EEG evaluation of suspected epilepsy can also be used in the evaluation of suspected pseudosyncope because hyperventilation and photic stimulation can provoke attacks. Another method used in the assessment of syncope is the tilt table test, which can be combined with EEG recording. It is important to remember that the general rule of medicine that diagnostic tests need to be interpreted in the context of the clinical presentation and history applies here as well. Comorbidities of functional and organic neurological disorders are common, and proving with EEG that a particular episode of apparent loss of consciousness was a functional attack does not rule out that the patient is suffering from epileptic seizures or syncopes as well.

5.2 Differentiation of functional from organic motor and sensory disorders

Neurophysiological techniques may also aid the diagnostic workup of other suspected psychogenic disorders. Although psychogenic sensory loss (blindness, deafness, anaesthesia) is very rare, psychogenic motor disorders (with paralysis or hyperkinetic features) are often encountered (in up to one third of patients) in specialist movement disorder clinics (Hallett et al. 2011). Intact function of the primary sensory or motor pathways can be assessed by evoked potentials (EPs) (see Chapter 3). A normal latency and amplitude of the pattern reversal VEP suggests intact processing up to the visual cortex, and the early and late components of the AEP can be used to assess the auditory pathways through the brainstem and up to auditory cortex, respectively. Similarly, intact processing of tactile afferents up to primary and secondary somatosensory cortex can be evaluated by SEPs. The motor system can be probed with MEPs, where single pulses of TMS over the motor cortex produce a response in peripheral muscles that can be recorded with electromyography (EMG) if the corticospinal tract is functional. In psychogenic motor or sensory disorders these electrophysiological parameters are generally normal, but later components of the event-related potential that are more strongly modulated by cognitive or affective processes can be altered (Vuilleumier 2009).

In hyperkinetic disorders, where a cortical (not necessarily voluntary) generation is suspected, demonstrating a readiness potential before the onset of the motor act can be diagnostically useful. This is particularly relevant in the differentiation of functional from propriospinal myoclonus because a spinal origin of myoclonus would not represent itself with a readiness potential. The increasing use of readiness potential recordings in these cases has revealed that the majority of patients previously classified as 'propriospinal myoclonus' in fact are suffering from a functional movement disorder (van der Salm et al. 2014).

5.3 Sleep EEG and behavioural disorders during sleep

The recording of physiological parameters during sleep is called polysomnography. It includes EEG, generally with a reduced number of electrodes (for example, a frontal, central, and occipital pair with two additional reference electrodes), ECG, electrooculography (EOG), and surface EMG of facial and leg muscles. In addition, several breathing parameters (airflow, chest movements, oxygen saturation) are recorded. The EEG and EOG patterns form the basis for the canonical classification of sleep into five stages: REM (rapid eye movement) sleep and the four non-REM (NREM) stages. During a regular eight-hour sleep we usually run through four to five sleep cycles, each containing non-REM sleep of varying depth, and towards the morning, also increasingly REM sleep (Fig. 5.2). Stage 1, light sleep, is characterized by theta activity on the EEG and reduced saccadic activity on the EOG. In stage 2, the EEG slows down further and shows characteristic sleep-related features: sleep spindles, brief bursts of beta activity, and K-complexes—single large waves that arise spontaneously or in response to external stimuli. The deep sleep stages 3 and 4 (collectively termed Stage N3 in the new terminology of the American Academy of Sleep Medicine) are characterized by the complete absence of eye or limb movement and delta waves in the EEG, which occur for 20–50% of the time (stage 3) or more than 50% of the time (stage 4).

REM sleep is characterized by waking EEG patterns, intermixed with irregular waves. The other components of the polysomnography reveal that breathing becomes faster and shallower and the heart rate increases. The EOG shows the eponymous rapid saccadic movements.

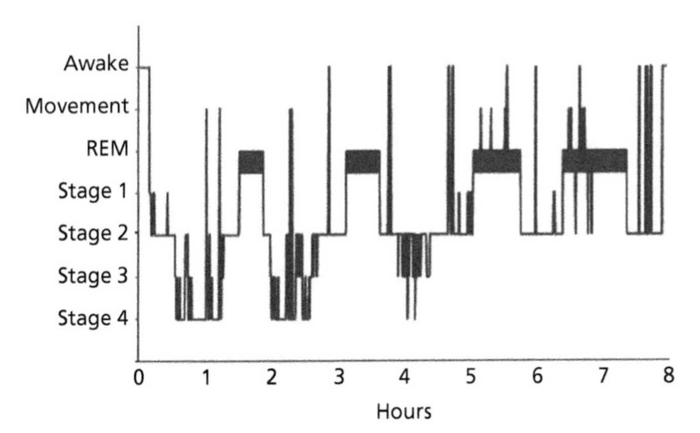

Fig. 5.2 A normal hypnogram (recording of physiological signals during sleep) showing the stages of sleep and their distribution over the night.
Reproduced from Wilson & Nutt, *Sleep Disorders*, 2013 with permission from Oxford University Press.

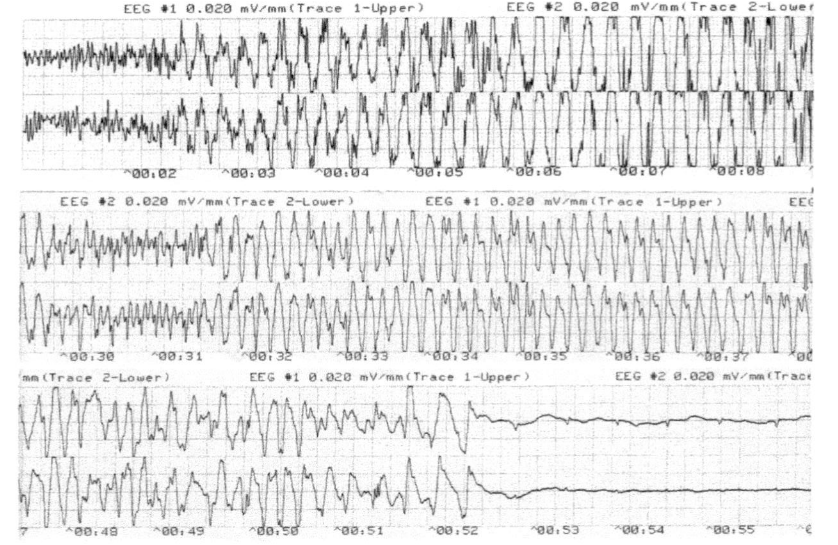

Fig. 5.3 EEG monitoring of ECT. This trace documents the successful induction of a generalized seizure of approx. 52 seconds in duration. The upper trace shows the build up towards the seizure with increased amplitude and slowing of the EEG. The middle trace includes the typical midictal epileptic activity, including spike/wave complexes and rhythmic high amplitude activity (starting between 00:31 and 00:32). The lower trace shows post-seizure suppression of the EEG (flat line from 00:53). The little spikes that can be seen on the flat line after the end of the seizure actually reflect the electrical activity of the heart (ECG recorded from the head).

Courtesy of Professor George Kirov.

Dreams mainly occur during REM sleep. This is also a period of generally reduced muscle tension, atonia, which prevents the dreams from being enacted. If the mechanisms that produce atonia fail, REM sleep behaviour disorder (RBD) can arise, which is one of the REM-sleep parasomnias. This condition is diagnosed on the basis of observer reports of dream enactment and EMG activity during REM sleep. It affects 0.5% of men over fifty years of age and is particularly prevalent in patients with Parkinson's disease.

Polysomnography is used in sleep and respiratory medicine for the workup of a variety of sleep-related disorders, which are beyond the remit of this book. It can also be helpful in the evaluation of nocturnal seizures. In the forensic-psychiatric context it can be useful for the evaluation of suspected sleep-related behavioural disorders, particularly sleepwalking and related NREM parasomnias. Several polysomnographic features have been described as being typical of sleepwalking– less slow wave (delta) activity than commonly observed during NREM sleep and arousal events that manifest themselves as disruptions of the sleep cycle or as 'cyclic alternating patterns', higher amplitude EEG waves with a periodicity of 20-40s (Cartwright and Guilleminault 2013). In rare cases in which an illicit act (for example, trespassing, stalking, or sexual acts) may be attributable to a NREM parasomnia polysomnography can be helpful in supporting such a diagnosis.

5.4 **Monitoring effectiveness of ECT**

Electroconvulsive therapy is mainly used in the treatment of refractory depression and as emergency treatment of febrile catatonia. It consists of the application of a strong electric pulse to the scalp (generally with bilateral electrode placement because this is more effective for the induction of seizures than unilateral application, albeit at higher risk of cognitive side effects) which induces a brief generalized seizure. Because ECT is conducted under muscle relaxation and anaesthesia, EEG monitoring is needed to document that a seizure has been successfully induced, and for how long it has lasted (Fig. 5.3). Most ECT machines have a two-channel EEG module to record seizure activity during the treatment. Failure to induce a seizure would be a reason for lack of a treatment effect. Whether the duration of the seizure or other electrophysiological parameters also influence the treatment's success is unknown, and more research is needed, both regarding the optimal administration of ECT and the biological mechanisms underlying its well-documented clinical effects.

References

Adachi N, Kanemoto K, de Toffol B, et al. (2013) Basic treatment principles for psychotic disorders in patients with epilepsy. *Epilepsia*, **54** Suppl 1:19–33.

Asadi-Pooya AA (2014) Transient epileptic amnesia: a concise review. *Epilepsy Behav*, **31**:243–5.

Elliott B, Joyce E, Shorvon S (2009) Delusions, illusions and hallucinations in epilepsy: 2. Complex phenomena and psychosis. *Epilepsy Res*, **85**(2–3):172–86.

Cartwright RD, Guilleminault C (2013) Defending sleepwalkers with science and an illustrative case. *J Clin Sleep Med*, **9**(7):721–6.

Fazel S, Lichtenstein P, Grann M, Långström N (2011) Risk of violent crime in individuals with epilepsy and traumatic brain injury: a 35-year Swedish population study. *PLoS Med*, **8**(12):e1001150.

Hallett M, et al. (2011) *Psychogenic movement disorders and other conversion disorders*. Cambridge: Cambridge University Press.

LaFrance WC, Baker GA, Duncan R, Goldstein LH, Reuber M (2013) Minimum requirements for the diagnosis of psychogenic nonepileptic seizures: a staged approach: a report from the International League Against Epilepsy Nonepileptic Seizures Task Force. *Epilepsia*, **54**(11):2005–18.

Linden DE (2014) *Brain Control*. Basingstoke: Palgrave Macmillan.

Mostacci B, Bisulli F, Alvisi L, Licchetta L, Baruzzi A, Tinuper P (2011) Ictal characteristics of psychogenic nonepileptic seizures: what we have learned from video/EEG recordings—a literature review. *Epilepsy Behav*,**22**(2):144–53.

Raj V, Rowe AA, Fleisch SB, Paranjape SY, Arain AM, Nicolson SE (2014) Psychogenic pseudosyncope: diagnosis and management. *Auton Neurosci*, **184**:66–72.

Shorter E, Healy D (2007) *Shock Therapy. The History of Electroconvulsive Treatment in Mental Illness*. New Brunswick, NJ: Rutgers University Press.

Treiman DM (1986) Epilepsy and violence: medical and legal issues. *Epilepsia*, **27** Suppl 2:S77–S104.

van der Salm SM, Erro R, Cordivari C, et al. (2014) Propriospinal myoclonus: Clinical reappraisal and review of literature. *Neurology*, **83**(20):1862–70.

Vuilleumier P (2009) The Neurophysiology of Self-Awareness Disorders in Conversion Hysteria. In Laureys S, Tononi G (eds.) *The neurology of consciousness*. London: Academic Press.

World Health Organization. *The ICD-10 classification of mental and behavioural disorders: clinical descriptions and diagnostic guidelines.* Available from: http://www.who.int/classifications/icd/en/bluebook.pdf [18 November 2015].

Zehtabchi S, Abdel Baki SG, Omurtag A, et al. (2013) Prevalence of non-convulsive seizure and other electroencephalographic abnormalities in ED patients with altered mental status. *Am J Emerg Med*, **31**(11):1578–82.

Chapter 6

Neuroimaging and diagnostic disease markers

Key points

- Radionuclide imaging markers of amyloid and tau protein deposition and altered brain metabolism may aid early detection of Alzheimer's disease-related pathology
- Structural brain imaging in schizophrenia and affective disorders has identified volume reductions in a variety of areas
- Neuroimaging and neurophysiology may aid the prediction of the course of (preclinical) mental illness, supported by the application of learning algorithms in data analysis
- Radionuclide imaging in schizophrenia corroborates a model of increased dopamine synthesis in the mesostriatal system, and in depression, shows some evidence of altered serotonin metabolism
- Many rare causes of neurodevelopmental disorders have characteristic neuroimaging features, but there are currently no imaging biomarkers for non-syndromic forms of autism, ADHD, or intellectual disability

One of the central aims of research in biological psychiatry is to provide biomarkers that can aid the diagnosis, classification, and prognosis of mental disorders. The definition of a biomarker is that it 'is objectively measured and evaluated as an indicator of normal biological processes, pathogenic processes, or pharmacologic responses to a therapeutic intervention' (Biomarker Definitions Working Group 2001). Psychiatry differs from other areas of medicine in that most of its disease categories are exclusively defined by the symptoms reported by patients and the behaviours they display, rather than any markers of (or presumptions about) the underlying biology. This has remained the case for the most recent diagnostic classification system, the DSM-5 of the American Psychiatric Association (American Psychiatric Association 2013), and presently there is no alternative to this approach. A more biologically grounded classification of diseases could result in higher reliability and validity of diagnoses and aid treatment stratification, and ultimately the discovery of new, rational treatments. It is important to remember here that 'biological' does not equal 'genetic'. Environmental influences on mental health also operate through biological processes and, assuming that their biological expression is fairly consistent across individuals, should be detectable in biomarker studies. One indication that it is possible, at least in principle, to detect the effects of environmental stressors in the brain comes from the large body of animal studies looking at the effects of stress on the volume and integrity of the hippocampus (McEwen 2007). Beyond diagnosis and classification, another important potential application of biomarkers is the early identification of people at risk of developing a mental illness. The most active research in this respect concerns the prediction of conversion from prodromal syndromes to schizophrenia and from mild cognitive impairment (MCI) to Alzheimer's disease (AD). The main purpose of such predictive biomarkers would be to enrich therapeutic trials for early intervention and prevention of these disorders with patients who, albeit still in preclinical stages, are at ultra-high risk and thus would benefit to the greatest extent from such interventions.

6.1 Dementia

Although dementia is a clinical diagnosis several clinical neuroimaging markers from MRI and PET are available to help the distinction between different organic causes of dementia such as AD or cerebrovascular disease (see Chapter 4). For example, FDG-PET distinguishes AD from other dementias (FTD, DLB) with high classification accuracy and may aid the prediction of future AD in people with MCI and even in cognitively normal individuals (Mosconi et al. 2010). FDG-PET assesses the effects of AD-related pathology on neural activity by picking up reductions in metabolism in areas of neuron loss. PET also provides an opportunity to probe the molecular pathology of AD directly, using tracers for amyloid or tau proteins. Tau-tracers are still under development, but 18F-THK5105 (Okamura et al. 2014) and other tracers have shown promise in indicating AD pathology in vivo. Increased levels of amyloid in the cerebral cortex can be demonstrated in most patients with manifest AD and in a high proportion (around 70%) of those with MCI, but also in some healthy elderly people (10–30%). Although this overlap between AD and MCI groups makes amyloid binding less useful as a diagnostic biomarker there is some evidence that the MCI patients with high amyloid burden are more likely to convert to AD than the amyloid-negative MCI group (Cohen and Klunk 2014).

Such a high-risk group (based on the combination of neuropsychological deficits in MCI and amyloid pathology) would be a very important target group for new treatment trials. Several recent high-profile trials of new AD therapeutics (passive immunization with monoclonal antibodies against amyloid) have failed to show clinical effects (Prins and Scheltens 2013), and the field is moving in the direction of early intervention. The approval of new therapeutics for this preclinical stage of dementia will crucially depend on the demonstration of preventative effects, which in turn requires trials with participants that have a high conversion risk in order to achieve sufficient power and tractable observation periods.

6.2 Psychotic disorders

6.2.1 Structural imaging

The enlargement of cerebral ventricles in patients with schizophrenia was one of the first replicated results of psychiatric imaging, and still constitutes one of the most robust imaging abnormalities associated with psychosis. However, it has neither the sensitivity nor the specificity needed for a clinical biomarker, even for the distinction between patients and unrelated controls, let alone for the distinction between patients with schizophrenia and bipolar disorder (who also have some degree of ventricular enlargement) or between patients and their unaffected relatives (McDonald et al. 2006). Corresponding to the ventricular enlargement, both overall white and grey matter

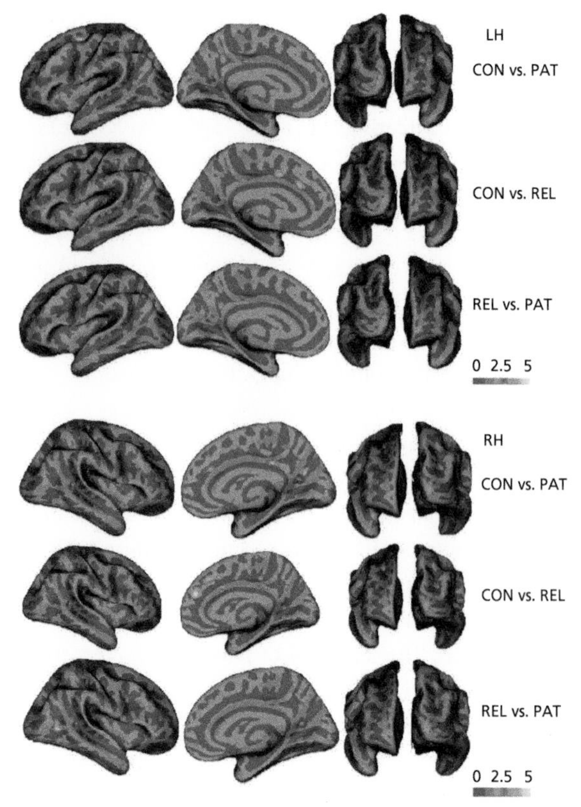

Fig. 6.1 Example of cortical thickness mapping in patients with schizophrenia, their unaffected relatives, and unrelated control participants. The upper three rows show group comparisons between controls and patients, controls and relatives, and relatives and patients in the left hemisphere; the lower three rows show group comparisons between controls and patients, controls and relatives, and relatives and patients in the right hemisphere. The colour bars denote t values (red to yellow: higher thickness in the first compared with the second group of each contrast). Several regions, particularly in the lateral frontal and temporal lobes and along the cingulate sulcus, showed reduced cortical thickness in patients and relatives compared to controls.

Reproduced from *Cereb Cortex*, 23,1, Oertel-Knöchel V, et al., 'Association between psychotic symptoms and cortical thickness reduction across the schizophrenia spectrum', PP. 61–70., Copyright (2013) with permission from Oxford University Press (Oertel-Knöchel et al. 2013).

are slightly reduced in schizophrenia. The reduction of grey matter progresses with the duration of illness, particularly in prefrontal and temporal cortex, and seems to be enhanced by atypical antipsychotics (Fusar-Poli et al. 2013 and Vita et al. 2012). However, even drug-naïve patients show volume reductions (albeit not as pronounced as treated patients), especially in the caudate nucleus and thalamus (Haijma et al. 2013).

In the temporal lobe, volume reduction generally affects the left more than the right hemisphere (Haijma et al. 2013). Such asymmetric brain changes have long been discussed as features of schizophrenia. In most healthy right-handed humans, the left planum temporale, the area on the surface of the temporal lobe behind the auditory cortex that comprises Wernicke's area, is larger on the left than on the right side, by approximately 20%. Several structural imaging studies have suggested that this temporal lobe asymmetry is reduced or even absent in patients with schizophrenia. Although the differences between patients (and in some studies, also between their first-degree relatives) and controls are not large and robust enough to produce a biomarker, they are of neuropsychological interest. The planum temporale asymmetry has been linked with the language dominance of the left hemisphere, and many of the core symptoms of schizophrenia are associated with abnormal language processing (auditory verbal hallucinations, thought disorder, neologisms, alogia). Schizophrenia is also associated with reduced or altered lateralization of cortical processing in other domains (Oertel-Knöchel and Linden 2011).

6.2.2 Neurophysiology

The most successful amongst the proposed electrophysiological indices of schizophrenia has been the mismatch negativity. This component of the auditory event-related potentials occurs approximately 200ms after a deviant stimulus (that differs from a series of stimuli in duration, pitch, intensity, or locus of origin) and indexes attentional processes. The mismatch negativity (MMN) is reduced in patients with schizophrenia compared to controls, but also in people at genetic risk. However, the MMN does not help the diagnostic differentiation between schizophrenia and other psychotic disorders because it is reduced in patients with bipolar disorder as well, which may point to a shared mechanism of impaired processing of changes in the environment across psychotic disorders. Another event-related potential (ERP) that has been widely studied across the psychotic spectrum is the P300 (see Chapter 3). It is similar to the MMN in that it indexes violations of sensory expectations, but it occurs slightly later in the chain of cortical processing (beyond 300ms). The P300 can be measured with oddball paradigms in all sensory modalities, and amplitude reductions have been most commonly reported for the auditory domain. Again, these changes are observed in attenuated form in at-risk individuals as well, but do not discriminate particularly well between patients with different psychotic disorders.

6.2.3 Imaging measures of neurochemistry

Radioligand imaging studies have provided some support for the dopamine model of schizophrenia. As treatment with antidopaminergic drugs can have a variety of effects on dopamine receptors and transporters and dopamine synthesis, studies in drug-naïve patients have the best validity. This group of patients has consistently shown evidence of mildly increased striatal dopamine synthesis, as measured by ^{18}F-DOPA PET (Fusar-Poli and Meyer-Lindenberg 2013). Data from drug-naïve, drug-free, and recently medicated patients taken together, thus excluding those with long-term use of dopamine blocking drugs, additionally show decreases in dopamine transporters and receptors in the striatum and cortex, which may be indirect evidence for tonically increased dopamine levels. Reductions of 5-HT1A and 5-HT2A receptors and GABA-A receptors have also been reported (Nikolaus et al. 2014). Magnetic resonance spectroscopy (MRS) studies focusing on GABA have also reported some changes in patients with schizophrenia compared to healthy controls, mainly lower GABA levels across several brain regions (Wijtenburg et al. 2015). This would conform to pathophysiological models that posit dysfunction of inhibitory circuits in schizophrenia (Lewis et al. 2005). However, MRS results on GABA in schizophrenia have been inconsistent, with several studies reporting no change. Furthermore, they can be influenced by benzodiazepine medication. MRS studies focusing on glutamate have reported elevated levels in the striatum in first episode patients, whereas patients in later stages of the disease were reported to have decreased glutamate levels across several brain regions (Wijtenburg et al. 2015). One caveat is that many of these measurements also include glutamine because this metabolite is difficult to differentiate from glutamate on MRS. Furthermore, one general limitation of MRS is that it does not distinguish between intra- and extracellular (and indeed between neuronal and glial) pools of metabolites, and thus it is difficult to infer functional properties of a neurotransmitter system from the observed concentration changes. Finally, none of the PET, SPECT or MRS findings in schizophrenia have achieved the status of a diagnostic biomarker.

6.2.4 Diagnostic and prognostic classification

Although there is no clear-cut imaging biomarker of schizophrenia, several studies have attempted to use pattern classification (see Chapter 2) of imaging data to differentiate patients with schizophrenia from healthy controls or to predict the conversion of prodromal patients to clinical schizophrenia. The conversion from the prodromal state of schizophrenia to the full clinical syndrome over a four-year follow-up may be predictable with 80% accuracy with whole-brain grey matter pattern analysis, even across imaging centres (Koutsouleris et al. 2014). Such multi-centre validation is particularly important for multivariate biomarkers because they depend strongly on the experimental parameters (e.g. numbers of scanned voxels) and analytical approaches (e.g. the algorithm used for feature selection). The clinical utility of such predictive markers will hinge on the question whether they add to the accuracy of prediction that can be achieved with conventional markers (such as clinical scales and neuropsychological tests).

6.3 Affective disorders

6.3.1 Structural imaging: unipolar disorder

Several hundred studies have investigated structural brain differences between patients with unipolar depression and healthy controls. Meta-analyses have consistently reported reduced volume of the hippocampus, but this may be confined to clinical episodes because patients in the remitted state do not seem to have reduced hippocampal volume. One interesting speculation is that antidepressant treatment may lead to a restitution of hippocampal volume, for example, by promoting neurogenesis as observed in animal studies, but longitudinal studies to support this idea are still lacking. Other areas of reduced volume are in the basal ganglia and the frontal lobe, and the cerebral ventricles are enlarged similar to the findings in several other psychiatric disorders (Kempton et al. 2011). It is worth remembering, though, that these differences are subtle (for example, in the range of 5% for hippocampal volume) and not detectable in individual patients.

6.3.2 Structural imaging: bipolar disorder

Structural changes in bipolar disorder have also been investigated in a large number of studies. A meta-analysis published in 2008 reported enlargement of the lateral ventricles (by approx. 17%) and widespread white matter hyperintensities as most consistent findings, but no specific regional grey matter reductions or increases (Kempton et al. 2008). A more recent meta-analysis, however, identified subtle frontal grey matter loss and volume increase of the globus pallidus in bipolar disorder (Arnone et al. 2009). Moreover, more recent data from a large multinational imaging study, the ENIGMA consortium, show subtle reductions of grey matter in the thalamus, amygdala, and hippocampus (Hibar et al. 2014).

It would be of considerable interest to identify patients with bipolar disorder early, already after the first depressive episode, because the long-term management would differ from that of patients with single or recurrent depressive episodes. So far no clinically predictive markers of bipolar disorder in patients with a depressive episode have been identified. Rates of deep white matter hyperintensities may be higher, and the corpus callosum smaller in bipolar disorder compared to unipolar depression (Kempton et al. 2011), but it is not known whether these features can serve as predictive markers early in the disease course.

6.3.3 Metabolic and neurochemical imaging

Although reports in the 1980s had indicated a 'hypofrontality' (particularly pronounced hypoactivity in the frontal lobes) and hemispheric asymmetries of frontal in activity in patients with depression, the comparison of regional cerebral blood flow (rCBF) or regional cerebral metabolic rate of glucose consumption (rCMRGlc) between acutely depressed patients and healthy controls across studies only yielded diffuse activity reduction in patients, but no specific spatial pattern (Nikolaus et al. 2000). Such diffuse hypometabolism during depressed states could, of course, be a correlate of the poverty of thought and general reduction in mental activity during the resting state during which such measures are acquired rather than a specific biomarker.

The most consistent result from PET and SPECT studies of the serotonergic system in depression has been a reduction of 5-HTT binding in the midbrain and amygdala (Gryglewski et al. 2014). Such a result may support a 'serotonin deficit' model of depression because 5-HTT density can be a proxy marker of serotonin concentrations. The apparent paradox that the most widely used antidepressants—the selective serotonin reuptake inhibitors (SSRI)—block a transporter whose density is already reduced in patients with depression, can be explained in the following way: if the reduction in 5-HTT indicates a primary deficit in serotonergic activity, blocking the 5-HTT even further would help re-balance this deficit by restoring the pool of synaptic 5-HT. At present, these results constitute an interesting lead but no definitive clue as to the validity of the 'monoamine hypothesis', which posits a primary deficit in synaptic serotonin and other monoamine neurotransmitters as a causal mechanism of depression.

Another neurotransmitter system that is increasingly implicated in the pathophysiology of major depressive disorder (MDD) is the GABAergic system of cortical interneurons. Some MRS studies have demonstrated decreases in cortical GABA concentrations in individuals with acute unipolar depression. However, there is insufficient evidence to suggest that GABA dysfunction is present during remission and would thus qualify as a potential trait marker of depression. Another limitation of GABA MRS in depression (and generally) is that data from the subcortical areas are difficult to obtain because of the large voxel sizes needed to obtain sufficient signal-to-noise (see Chapter 2) (Keedwell and Linden 2013).

6.4 Developmental disorders

6.4.1 Learning disabilities

Amongst the developmental disorders, the most salient brain imaging changes are probably observed in patients with learning disabilities ('mental retardation' in the ICD-10; 'intellectual disability' in the DSM-5). Learning disability is defined by an IQ below 70, which corresponds to two standard deviations below the population mean, and thus by definition affects about 2.5% of the population. Although this is a somewhat arbitrary cut-off point it seems to correspond to a level of functioning below which educational and employment prospects are severely diminished, and those affected often require psychosocial support. Generally, lower intelligence is associated with lower grey matter volume, although the shared variance is only around 10%. Common causes of learning disability include genetic syndromes (although each single syndrome is relatively rare) and perinatal brain damage, although in the majority of cases of mild learning disability the cause remains currently unknown. The contribution of genetic syndromes as well as the prevalence of brain abnormalities detected by imaging increase with the severity of the learning disability, and the detectability of genetic causes, and imaging correlates is likely to increase with more widespread and detailed genotyping and advances in imaging technology (particularly higher field strengths). So far CT and MRI studies have reported brain abnormalities in up to 60% of study participants with learning disabilities (Gothelf et al. 2005), although there is considerable variability across studies and not all of these abnormalities would be deemed to be clinically significant.

6.4.2 Malformations of cortical development

Malformations of cortical development (MCD) result from disturbed neuronal migration and postmigrational cortical organization, and are often associated with abnormal cell proliferation leading to micro- or megalencephaly. Cerebral folding (gyration) is frequently affected. MCD entities may also be defined by additional structural cerebral anomalies, alterations of the corpus callosum, basal ganglia, cerebellum or brain stem, or intracerebral calcification. The diagnosis is made on neuroimaging and/or histopathothology. Many of these malformations are caused by *de-novo* mutations affecting genes governing cortical development. MCD are usually associated with mild to severe developmental delay and intellectual disability, epilepsy, and other neurological deficits. Examples include polymicrogyria (more numerous and smaller gyri than normal), classical lissencephaly (absent gyri (agyria); wider and fewer gyri than normal (pachygyria)), subcortical band heterotopia (bands of grey matter in the subcortical white matter), and periventricular nodular heterotopias (nodules of grey matter adjacent to the wall of the lateral ventricles). Polymicrogyria (PMG) is the most common MCD, and is a very heterogeneous disorder (see Fig. 6.2D). Causes include intrauterine infections (e.g., cytomegalovirus), chromosome anomalies (e.g., 22q11.2 deletion), and intragenic mutations (*WDR62*). Classical lissencephaly (LIS), the paradigm of a neuronal migration disorder, is predominantly caused by mutations in the *LIS1* and *DCX* genes (see Fig. 6.2B). Subcortical band heterotopia (SBH) in females is commonly associated with mutations in the X-linked *DCX* gene (see Fig. 6.2C). Periventricular heterotopia (PH) may present as single or multiple nodules lining the ventricular walls. Bilateral, almost confluent nodules along the lateral ventricles are typically seen in females with mutation in the *FLNA* gene, also located on the X-chromosome (see Fig. 6.2A). Neuropsychiatric disorders, including anxiety, depression, schizophrenia, and autism have been reported in a number of patients, including females with *FLNA* mutations (Fry et al. 2013), suggesting that neuropsychiatric disease may be a potential consequence of PH.

6.4.3 **Genetic syndromes**

Learning disability and a wide range of developmental psychopathology are also associated with several genetic syndromes which do not have pathognomonic imaging correlates, although they may co-occur with specific brain abnormalities. The commonest genetic cause of learning disability (about 1/1000 births) is Down syndrome, caused by partial or complete trisomy of chromosome 21. Although brain scans may be normal on clinical inspection, quantitative studies have yielded consistent evidence for regional deficits in brain volume, particularly in the cerebellum, hippocampus, and frontal lobe. Down

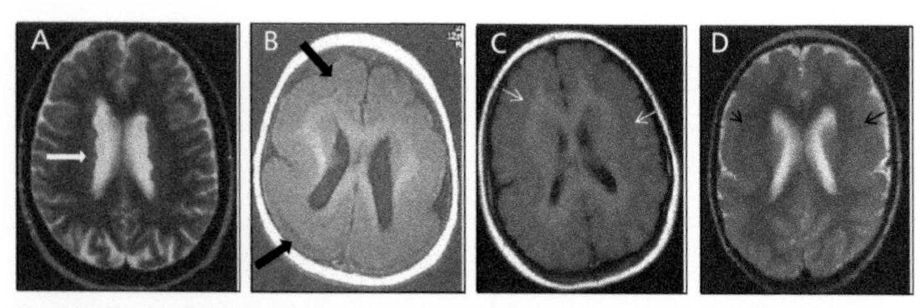

Fig. 6.2 Examples of MCDs. Axial T2 (A+D) and T1 (B+C) weighted MR images. A: bilateral periventricular nodular heterotopia (white arrow); B: classical lissencephaly with frontal pachygria and posterior agyria (black arrows); C: bilateral subcortical band heterotopia (white arrows); D: polymicrogyria (black arrows).

Courtesy of Professor Daniela Pilz.

syndrome is also associated with a higher occurrence of calcifications of the basal ganglia. In addition, older people with Down syndrome have a high load of amyloid depositions in the brain, probably because they have an extra copy of the gene for amyloid precursor protein (APP), and about half of them develop clinical signs of Alzheimer's disease. Correspondingly, older people with Down syndrome show evidence of amyloid binding on PET imaging, although the evidence base is still rather small.

A relatively common inherited cause of learning disability is the X-chromosome-linked fragile-X syndrome (FXS), which affects about 1/3000 males and 1/6000 females. The affected gene codes for the FMR1 (FXS and mental retardation 1) protein, which has been implicated in various stages of synaptic plasticity and neural development. Early brain overgrowth (EBO), which is detectable by age two, is one of the characteristic features of altered brain development in FXS. The basal ganglia and thalamus are particularly affected by this early excessive growth. Conversely, the cerebellar vermis is sometimes hypoplastic (Schaer and Eliez 2007).

Individuals with velocardiofacial syndrome (VCFS), another relatively common genetic cause of (generally mild) learning disability that furthermore confers a high (approx. 25%) risk of developing a psychotic disorder, have a deletion of part of the long arm of chromosome 22 (in the 22q11.2 region). VCFS—also termed DiGeorge syndrome in recognition of the paediatrician Angelo diGeorge who first described it in 1968—is characterized by a combination of physical defects, for example, cleft palate, cardiac abnormalities, reduced immune function (because of aplasia of the thymus), and skeletal abnormalities. Sufferers also have a higher incidence of epilepsy. VCFS affects about 1/4000 live births. Individual brain imaging may not bring up any abnormality, but comparisons between groups of VCFS sufferers and control participants have revealed reduced grey and white matter volumes, particularly in posterior areas of the brain. This relative deficit in the development of the parietal lobe may explain the specific visuospatial and arithmetic deficits characteristic of the neuropsychological profile of VCFS. Some evidence suggests that those who go on to develop schizophrenia also develop a volume reduction in the frontal lobes. The volume of the cerebellar vermis is also reduced. Conversely, the lateral ventricles are increased in volume (Schaer and Eliez 2007). VCFS is also associated with a higher occurrence of abnormalities that can be detected on individual scans such as cysts around the frontal horns of the lateral ventricles or polymicrogyria. Williams syndrome (or Williams-Beuren syndrome) is another, rarer genetic cause of learning disability and multiple physical problems, associated with a microdeletion at 7q11.23. Deficits in visuospatial processing are prominent, like in VCFS, and here, too, volume reductions generally affect the posterior part more than the anterior part of the brain. The development of sequencing technology over the last decade has revealed several more microdeletion and microduplication syndromes (collectively termed 'copy number variants' or CNVs) that confer substantial risk of developmental delay, epilepsy, and psychopathology. Imaging studies of these syndromes are only just starting to map out the associated brain abnormalities (Stefansson et al. 2014), and for most CNVs it is unknown which of the affected genes contribute to the alterations in brain development.

Imaging findings in neurodevelopmental disorders without learning disability have generally been less salient. Although there is some preliminary evidence for reduced basal ganglia volumes in attention deficit hyperactivity disorder (ADHD), most affected children will not show any abnormalities on clinical imaging. One interesting imaging feature of autism spectrum disorders has been EBO followed by premature arrest of brain growth (Courchesne et al. 2007), similar to the pattern observed in FXS, although most patients scanned during adolescence or adulthood will have normal imaging findings.

References

American Psychiatric Association (2013) *Diagnostic and Statistical Manual of Mental Disorders: DSM-5,* 5th edn. Washington, DC: American Psychiatric Publishing.

Arnone D, Cavanagh J, Gerber D, Lawrie SM, Ebmeier KP, McIntosh AM (2009) Magnetic resonance imaging studies in bipolar disorder and schizophrenia: meta-analysis. *Br J Psychiatry*, **195**(3):194–201.

Biomarkers Definitions Working Group (2001) Biomarkers and surrogate endpoints: preferred definitions and conceptual framework. *Clin Pharmacol Ther*, **69**(3):89–95.

Cohen AD, Klunk WE (2014) Early detection of Alzheimer's disease using PiB and FDG PET. *Neurobiol Dis*, **72** (Pt A):117–22.

Courchesne E, Pierce K, Schumann CM, et al. (2007) Mapping early brain development in autism. *Neuron*, **56**(2):399–413

Fry AE, Kerr MP, Gibbon F, et al. (2013) Neuropsychiatric disease in patients with periventricular heterotopia. *J Neuropsychiatry Clin Neurosci*, **25**(1):26–31.

Fusar-Poli P, Meyer-Lindenberg A (2013) Striatal presynaptic dopamine in schizophrenia, part II: meta-analysis of [(18)F/(11)C]-DOPA PET studies. *Schizophr Bull*, **39**(1):33–42.

Fusar-Poli P, Smieskova R, Kempton MJ, Ho BC, Andreasen NC, Borgwardt S (2013) Progressive brain changes in schizophrenia related to antipsychotic treatment? A meta-analysis of longitudinal MRI studies. *Neurosci Biobehav Rev*, **37**(8):1680–91.

Gothelf D, Furfaro JA, Penniman LC, Glover GH, Reiss AL (2005) The contribution of novel brain imaging techniques to understanding the neurobiology of mental retardation and developmental disabilities. *Ment Retard Dev Disabil Res Rev*, **11**(4):331–9.

Gryglewski G, Lanzenberger R, Kranz GS, Cumming P (2014) Meta-analysis of molecular imaging of serotonin transporters in major depression. *J Cereb Blood Flow Metab*, **34**(7):1096–103.

Haijma SV, Van Haren N, Cahn W, Koolschijn PC, Hulshoff Pol HE, Kahn RS (2013) Brain volumes in schizophrenia: a meta-analysis in over 18 000 subjects. *Schizophr Bull*, **39**(5):1129–38.

Hibar D, Westlye L, Thompson P, Andreassen O, ENIGMA Bipolar Disorder Working Group (2014) ENIGMA Bipolar disorder working group findings from 1,747 cases and 2,615 controls. Organisation for Human Brain Mapping 2014 Annual Meeting, abstract 1363.

Keedwell PA, Linden DE (2013) Integrative neuroimaging in mood disorders. *Curr Opin Psychiatry*, **26**(1):27–32.

Kempton MJ, Geddes JR, Ettinger U, Williams SC, Grasby PM (2008) Meta-analysis, database, and meta-regression of 98 structural imaging studies in bipolar disorder. *Arch Gen Psychiatry*, **65**(9):1017–32.

Kempton MJ, Salvador Z, Munafò MR, et al. (2011) Structural neuroimaging studies in major depressive disorder. Meta-analysis and comparison with bipolar disorder. *Arch Gen Psychiatry*, **68**(7):675–90.

Koutsouleris N, Riecher-Rössler A, Meisenzahl EM, et al. (2014) Detecting the Psychosis Prodrome Across High-risk Populations Using Neuroanatomical Biomarkers. *Schizophr Bull*, **10**(1093).

Lewis D, Hashimoto T, Volk D (2005) Cortical inhibitory neurons and schizophrenia. *Nat Rev Neurosci*. **6**(4):312–24.

McDonald C, Marshall N, Sham P, et al. (2006) Regional brain morphometry in patients with schizophrenia or bipolar disorder and their unaffected relatives. *Am J Psychiatry*, **163**(3):478–87.

McEwen BS (2007) Physiology and neurobiology of stress and adaptation: central role of the brain. *Physiol Rev*, **87**(3):873–904.

Mosconi L, Berti V, Glodzik L, Pupi A, De Santi S, de Leon MJ (2010) Pre-clinical detection of Alzheimer's disease using FDG-PET, with or without amyloid imaging. *J Alzheimers Dis*,**20**(3):843–54.

Nikolaus S, Hautzel H, Müller HW (2014) Neurochemical dysfunction in treated and nontreated schizophrenia—a retrospective analysis of in vivo imaging studies. *Rev Neurosci*, **25**(1):25–96.

Nikolaus S, Larisch R, Beu M, Vosberg H, Müller-Gärtner HW (2000) Diffuse cortical reduction of neuronal activity in unipolar major depression: a retrospective analysis of 337 patients and 321 controls. *Nucl Med Commun*, **21**(12):1119–25.

Oertel-Knöchel V, Knöchel C, Rotarska-Jagiela A, et al. (2013) Association between psychotic symptoms and cortical thickness reduction across the schizophrenia spectrum. *Cereb Cortex*, **23**(1):61–70.

Oertel-Knöchel V, Linden DE (2011) Cerebral asymmetry in schizophrenia. *Neuroscientist*, **17**(5):456–67.

Okamura N, Furumoto S, Fodero-Tavoletti MT, et al. (2014) Non-invasive assessment of Alzheimer's disease neurofibrillary pathology using 18F-THK5105 PET. *Brain*, **137**(Pt 6):1762–71.

Prins ND, Scheltens P (2013) Treating Alzheimer's disease with monoclonal antibodies: current status and outlook for the future. *Alzheimers Res Ther*, **5**(6):56.

Schaer M, Eliez S (2007) From genes to brain: understanding brain development in neurogenetic disorders using neuroimaging techniques. *Child Adolesc Psychiatr Clin N Am*, **16**(3):557–79.

Stefansson H, Meyer-Lindenberg A, Steinberg S, et al. (2014) CNVs conferring risk of autism or schizophrenia affect cognition in controls. *Nature*, **505**(7483):361–6.

Vita A, De Peri L, Deste G, Sacchetti E (2012) Progressive loss of cortical gray matter in schizophrenia: a meta-analysis and meta-regression of longitudinal MRI studies. *Transl Psychiatry,* **2**:e190.

Wijtenburg SA, Yang S, Fischer BA, Rowland LM (2015) In vivo assessment of neurotransmitters and modulators with magnetic resonance spectroscopy: Application to schizophrenia. *Neurosci Biobehav Rev*, **51**:276–95.

Chapter 7

Neuroimaging and mechanisms of mental disorders

Key points

- Imaging can support neuropsychological disease models of mental disorder by demonstrating altered structure and/or function in specific neural circuits and during specific cognitive tasks
- Paradigms investigating domains of altered cognitive, motivational, or autonomic functions (research domain criteria or RDoC) have been proposed
- Biological mechanisms of 'model psychoses' (induced with medication) can be studied with neuroimaging and neurophysiology
- Many neuropsychiatric disorders have a high heritability, with contributions from rare variants with high penetrance and common variants with low penetrance
- Mechanisms of risk genes can be studied in unaffected carriers of the risk variant without the confounds commonly present in patient populations

7.1 Neuropsychological disease models

7.1.1 From hypofrontality to impaired temporal difference learning

Although many aspects of the pathophysiology of Alzheimer's disease (AD) are still unclear we can at least put forward a neuropsychological disease model that links the affected brain areas with the cognitive and behavioural impairment. For example, the early damage to the hippocampus has been implicated in the problems with memory encoding. In the absence of characteristic focal brain changes, it is much less straightforward to construe similar models about psychotic, affective, or anxiety disorders. When it became possible to assess regional cerebral blood flow and metabolism in patients with schizophrenia from the 1960s onwards, researchers initially focused on the concept of 'hypofrontality', which implied that an underactive frontal lobe explained some of the clinical and cognitive symptoms of schizophrenia such as impaired working memory and executive functions. Although hypofrontality was reported by many studies using a variety of techniques during both rest and activation studies there is considerable overlap in the activation patterns between patients and healthy controls. Furthermore, fMRI studies in particular have produced evidence for both hypo- and hyperactivation during cognitive tasks, suggesting that the association between regional activation levels and task performance in schizophrenia is more complex. For example, higher activation levels in patients compared to controls may indicate less efficient neural processing, or compensatory activity in the face of higher subjective task difficulty.

An imbalance between frontal (too little) and limbic (too much) activation has been suggested to underlie some of the emotion regulation difficulties of patients with affective disorders. However, here too, the evidence has been mixed, and more recent research suggests that functional subdivisions of the frontal lobe, for example, into a more lateral 'voluntary' and medial 'automatic' emotion regulation system, need to be taken into account (Rive et al. 2013).

Over the last decade, computational neuroscience has become a very influential framework of functional activation studies in schizophrenia and beyond. One example of this approach is the combination of computational modelling of learning rates in temporal difference reward learning (TDRL) paradigms with fMRI. TDRL models assume that humans adjust their behaviour to maximize rewards by minimizing the 'reward prediction error', and the discrepancy between the anticipated and the actual reward. Animal neurophysiology suggests that the prediction error is encoded by the dopaminergic system, which adds to its potential interest for the study of psychiatric disorders. For example, altered behavioural and neural responsiveness to reward has been found in depression (reduced reward sensitivity) and addiction (increase in the reward expected from the substance of abuse). These behavioural and neural learning models can thus explain some of the behavioural features of the disorder. Along similar lines, the delusions of psychotic disorders have been associated with faulty prediction error signalling (Wang and Krystal 2014).

7.1.2 RDoC: a recent innovation

One of the reasons it has been so difficult to establish consistent neuropsychological disease models of psychiatric disorders may be their phenotypic heterogeneity. The diagnostic categories comprise people who differ in many behavioural and emotional domains; for example, there are schizophrenia patients with low and high impulsivity or low and high needs for close relationships. In order to address this problem the National Institute for Mental Health (NIMH) in the US has developed a new system that allows for the classification of mental disorders along dimensions of behaviour rather than along the conventional diagnostic categories. It is hoped that these RDoC (see Box 7.1) will provide more consistent neurobiological underpinnings and reliable relationships between neural and behavioural phenotypes.

Although for the time being the RDoC are mainly intended as a framework for research into the biological processes underlying dysfunctional behaviours, cognitions, and emotions, they could ultimately lead to a new diagnostic system where the current categorical taxonomy of well-demarcated diseases is replaced by a dimensional approach. For example, rather than being classified as suffering from major depression, a patient might be described as suffering from heightened sensitivity to frustrative non-reward and lower approach motivation and responsiveness to reward attainment. Although such a dimensional approach may initially seem to be more complex than the current diagnostic algorithms, it could facilitate more specific treatments—after all the negative and positive valence systems may operate through partly different neurotransmitter systems (e.g. predominantly serotonin versus dopamine, although this is still speculative) and thus be modulated by different drugs. One major challenge is that, unlike DSM or ICD, the RDoC do not yet incorporate the longitudinal course and often episodic nature of mental disorders (Owen 2014).

Box 7.1 The RDoC proposed by the NIMH to facilitate a classification of mental disorders that is closer to the underlying biology

◆ **Negative valence systems (responses to aversive situations, avoidance behaviour)**
Responses to acute threat (fear)
Responses to potential harm (anxiety)
Responses to sustained threat
Frustrative non-reward
Loss
◆ **Positive valence systems (responses to potential reward, approach behaviour)**
Approach motivation
Reward valuation
Effort valuation/willingness to work
Expectancy/reward prediction error
Action selection/preference-based decision-making
Initial responsiveness to reward attainment
Sustained/longer-term responsiveness to reward attainment
Reward learning
Habit
◆ **Cognitive systems**
Attention
Perception
Declarative memory
Language
Cognitive control
Working memory
◆ **Systems for social processes**
Affiliation and attachment
Social communication: reception/production of facial/non-facial communication

Perception and understanding of self: agency/self-knowledge
Perception and understanding of others: animacy perception/action perception/understanding mental states
◆ **Arousal/regulatory systems**
Arousal
Circadian rhythms
Sleep and wakefulness

Adapted from *NIMH Research Domain Criteria (RDoC) Matrix*. Accessed 10 August 2015, from http://www.nimh.nih.gov/research-priorities/rdoc/research-domain-criteria-matrix.shtml and subordinate webpages.

Note: The RDoC project is a dynamic, evolving process, and as such, the matrix is subject to continual updates and changes in presentation and format; for the most up-to-date and accurate version, visit the website noted.

7.1.3 **Neuroimaging and RDoC**

Neuroimaging can play a major role in the validation of the RDoC approach. Twenty years of functional neuroimaging research, mainly with fMRI but also with PET, MEG, and EEG, have created a substantial body of knowledge about the neural systems underpinning the domains and subdomains of perception, affect, learning, and cognition in humans. For example, threat responses (part of the RDoC 'negative valence systems') are encoded in the limbic system, particularly the amygdala. Regarding the RDoC 'positive valence systems', reward-based learning operates through mesocortico-(ventral) striatal pathways, whereas habit formation seems to involve activity of fronto-(dorsal) striatal pathways. The largest body of fMRI studies has probably been conducted on the domain of the RDoC 'cognitive systems', revealing networks subserving attention (fronto-parietal), declarative memory (frontal-mesiotemporal), working memory (fronto-parietal), language (inferior frontal-superior temporal), and cognitive control (several subsystems in frontal cortex). Aspects of 'social communication' have also been mapped onto hubs in the brain, for example, the posterior superior temporal gyrus for analysis of facial expression or the medial part of the frontal lobe for self-knowledge. There is also a considerable body of literature on a network subserving 'theory of mind', which corresponds to the RDoC 'understanding mental states', which includes medial frontal regions, the posterior cingulate gyrus, and temporoparietal junction bilaterally, areas that are also found to be active during the 'idling' process of the brain and have therefore been termed a 'default-mode network'. Although the 'arousal/regulatory systems' are harder to access by fMRI because of their location in the brainstem and diencephalon they have characteristic EEG signatures, as discussed in Chapter 5.

It remains to be seen whether RDoC are clinically useful— that is, if they have good reliability and correlate with subjective and objective impairments and, at least to some extent, with treatment responses. The next step would then be to understand how and why disturbances along the RDoC dimensions arise. This mechanistic understanding requires an integrated translational neuroscience approach that incorporates both animal and human studies. For example, if longitudinal human imaging studies can elucidate the neural deficits associated with the emergence of dysfunctional affiliative behaviour, and more detailed neuroanatomical and neurophysiological workup in rodent models then identifies the specific neurochemical and neuroendocrinological changes, this could lead to a completely new mechanistic understanding of some developmental disorders and to new, rational pharmacological treatments.

7.2 **Pharmacological disease models**

The mechanisms of psychiatric disorders can also be elucidated through the experimental induction of specific symptoms or syndromes (see Table 7.1). For example, so-called 'model psychoses' can be induced with ketamine, phencyclidine (PCP), amphetamines, and hallucinogens such as lysergic acid diethylamide (LSD) or psilocybin. A similar approach is possible for the experimental induction of transient anxiety. Panic attacks can be induced through the administration of fragments of the neuropeptide cholecystokinin (CCK).

These pharmacological models for psychiatric disorders reveal a biological mechanism that can underlie psychotic or anxiety syndromes. This information can be used to introduce new experimental treatments. For example, the glutamate model of schizophrenia, which is supported by the ketamine studies (Poels et al. 2014), has given rise to new treatment approaches with NMDA-receptor modulators such

Table 7.1 Main imaging findings for drug-induced disease models

Substance class	Main neurotransmitter systems	Psychological effects	fMRI/PET/MRS effects	MEG/EEG effects
Hallucinogens	5-HT2A/1A receptor agonists	Perceptual changes, synaesthesia, visual hallucinations, flashbacks	Increased prefrontal glucose metabolism	Reduced N170 (ERP component evoked by faces)
Amphetamine	Dopamine reuptake inhibition	Activation, euphoria, aggression	Decreased binding of D2/D3 receptor ligands, indicating enhanced dopaminergic transmission in striatum and cortex	Reduced latency and increased amplitude of auditory P300 (mixed results)
Ketamine, PCP	NMDA receptor antagonist, D2 receptor agonist	Dissociation, hallucinations, analgesia	Decreased binding of D2/D3 receptor ligands, indicating enhanced dopaminergic transmission in striatum and some cortical areas (mixed results); increased cortical glutamate concentrations	Reduced P1 (component of visual EP), MMN
Cholecystokinin (CCK)-tetrapeptide	CCK system (peptide hormone)	Panic attacks	Increased blood flow in cingulate cortex and other limbic and paralimbic areas; increased glutamate concentration in ACC	

as D-serine, although definitive evidence for its efficacy is still missing. One limitation of this approach is that even if a schizophrenia-like psychosis can be consistently induced with an NMDA-receptor antagonist this does not imply that this mechanism is involved in the majority of patients with schizophrenia. Only a small proportion might suffer from such a 'glutamate deficit' psychosis and identifying these patients for trials of modulators of the glutamate system would be paramount. One of the most active areas of psychiatric imaging research currently evolves around the need for biomarkers for such treatment stratification.

7.3 Genetic disease models

The identification of risk genes for mental disorders has progressed considerably over the last two decades. Although the high heritability of autism, schizophrenia, and bipolar disorder, in particular, had been known for a long time, only a few, rare risk variants had been known until the mid-1990s. Subsequent work has resulted in an expansion of the number of known rare, highly penetrant copy number variants (CNVs) associated with neurodevelopmental disorders and revealed the first common risk loci with genome-wide significance.

7.3.1 Imaging rare variants

CNVs are deletions or duplications of segments of chromosomes, ranging from about a kilobase to several megabases in length. They can be inherited or occur *de novo*. CNVs with a high penetrance (high associated disease risk) are rare, although their prevalence and penetrance are not precisely known because genotyping studies in large population cohorts are only just beginning. Studies of CNVs in schizophrenia have implicated a number of specific chromosomal loci including 1q21.1, 3q29, 15q11.2, 15q13.3, 16p11.2, 16p13.1, and 22q11.2, with estimated odds ratios ranging between two and 60. Imaging studies of carriers of CNVs offer the opportunity to investigate the effects of genetic variation on brain structure and function in a biologically homogenous at-risk group.

22q11.2 deletion or velocardiofacial syndrome (VCFS) (see Chapter 6) is one of the most penetrant CNVs associated with schizophrenia, and the variant that has been studied most extensively in imaging studies. The 22q11.2 deletion is also associated with other neurodevelopmental manifestations including cognitive impairment, autism, and ADHD and with a wide range of physical abnormalities. This phenomenon, where a genetic variant can lead to a range of phenotypic features, is called pleiotropy.

MRI studies have identified both qualitative and quantitative differences between patients with VCFS and controls. Qualitative features on MRI scans are those that can be identified on individual scans, like in classical clinical imaging, although not all qualitative changes are clinically relevant. For example, VCFS is associated with an increased prevalence of polymicrogyria (see Chapter 6). These qualitative changes suggest that early cortical development and neuronal migration may be disrupted as a consequence of the deletion, although it is still unknown which of the over 40 affected genes are responsible.

Genetic risk models have been particularly influential in the exploration of the biology of AD because of the small fraction of cases that are caused by autosomal dominant mutations in the presenilin (PSEN) 1 and 2 or amyloid precursor protein (APP) gene. Furthermore, a common variant on the apolipoprotein E (APOE) gene (epsilon 4 allele with a frequency of about 20% in the general population) confers a notable risk of developing AD, making it the most penetrant common risk factor for any neuropsychiatric disorder (see Fig. 7.1). Carriers of the dominant mutations and the APOE epsilon 4 allele have been imaged with MRI and PET to test whether AD-related pathology is already present in the preclinical stages of the disease.

The AD-related mutations in the PSEN and APP genes are relevant for the amyloid pathology because they can lead to an increase in amyloid-beta formation and subsequently amyloid plaque formation (Linden 2012a). PET studies with amyloid tracers have revealed amyloid deposition in several brain areas including the mesial temporal lobe and posterior cingulate gyrus in asymptomatic carriers of the E280A mutations on the presenilin 1 gene already during the fourth decade of life, a decade or more before the onset of clinical dementia and also long before such depositions become apparent in unaffected elderly people. This group of unaffected mutation carriers also showed higher hippocampal activation (measured with fMRI) during a memory task than non-carriers, which may be interpreted as indicating

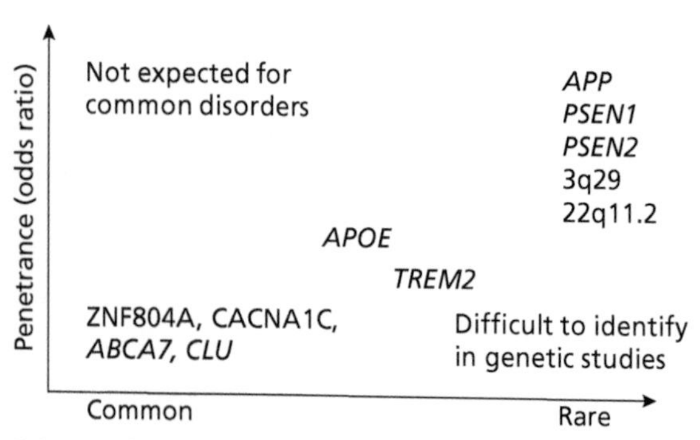

Fig. 7.1 Schema of the genetic architecture of neuropsychiatric (and other common) disorders with examples of risk loci for schizophrenia (normal font) and AD (italics). *ZNF804, CACNA1C, CLU,* and *ABCA7* are genes carrying common single nucleotide polymorphisms (SNPs) that have been identified as risk loci by genome-wide association studies. A rare missense mutation in *TREM2* increases risk of AD. This gene can also be affected by a homozygous loss-of-function mutation resulting in an autosomal recessive form of early-onset dementia. For further explanations see text.

compensatory mechanisms or higher effort needed to attain similar levels of cognitive performance (Fleisher et al. 2012 and Reiman et al. 2012). Similar findings of putative hyperactivation have been obtained in patients at the early stages of AD and other at-risk groups, including carriers of the APOE e4 allele and other common variants (Prvulovic et al. 2005) (see Section 7.3.2). In general it has been hypothesized that the progression of AD pathology is characterized by changes in brain chemistry (deposition of dysfunctional amyloid and tau proteins) and structure that starts decades before the clinical manifestation of dementia (see Fig. 7.1) (Jack et al. 2010). Early detection of this AD pathology through imaging or chemical analysis of blood or CSF is a key aim of neuropsychiatric biomarker research.

7.3.2 Imaging common variants and polygenic risk

Although a small proportion (probably a few percent) of cases of schizophrenia have a major contribution from a single, highly penetrant variant, the majority of patients carry a high number of common risk variants which individually have low odds ratios. There may be thousands of such loci, with considerable genetic heterogeneity across patients. This implies that variation in many different genes can lead to similar phenotypes. This phenomenon is called convergence, and represents the converse scenario to pleiotropy. Such polygenic models probably apply to many common disorders with non-Mendelian inheritance, for example, AD, diabetes, and rheumatoid arthritis.

Common (as defined by a minor allele frequency >1%) single-base pair variants, SNPs, occur at several tens of millions of loci in the human genome. Because of the small odds ratios of individual variants and the large number of statistical tests (hundreds of thousands to millions, depending on the density of the array used for the genetic testing) large numbers of patients and control participants are needed to identify genome-wide significant risk alleles. Several genome-wide association studies (GWAS) have now identified common risk variants for schizophrenia, bipolar disorder, and AD. This method for the identification of risk variants is more robust than the previous candidate gene approach, where individual SNPs were selected based on their biological function, but subsequently failed to survive genome-wide significance thresholds and replication studies. However, the results are difficult to interpret because the molecular effects associated with the identified SNPs are often not known.

Understanding the effects of these variants on the brain will likely reveal more about the causal mechanisms of schizophrenia. Psychiatric genetics may also help identify subgroups of patients who need to be diagnosed and treated in different ways (stratification). However, very little is yet known about the effects of the common risk variants on brain and behaviour and how they contribute to clinical symptoms. Neuroimaging studies in carriers of these risk variants can help fill this gap of knowledge. For example, studies of healthy carriers of risk variants for schizophrenia or bipolar disorder have shown alterations in frontal and hippocampal function, similar to those sometimes observed in patient groups. In a similar vein, imaging studies of the carriers of the AD-risk variant on the clusterin (Apolipoprotein J) gene have revealed hyperactivation during cognitive tasks (Lancaster et al. 2014), similar to findings in carriers of Mendelian mutations (see Section 7.3.1) or patients in the early stages of the disease.

Although genetic imaging has brought up several associations between single SNPs and neural phenotypes only the minority of effects has been replicated, and the conventional sample sizes of neuroimaging studies (up to several hundred participants) may not be large enough robustly to detect the putatively small effect of single variants with low penetrance for the clinical phenotype. A more powerful approach to detect effects of common risk variants entails the creation of cumulative genetic risk scores (based on odds ratios from a database of highly significant SNPs). These polygenic risk scores can explain up to 25% of disease risk (for schizophrenia) and are thus likely to be strongly associated with any biological phenotypes of the disease as well. The first studies with this approach have revealed similar changes in frontal cortex activation and brain structure in healthy individuals with high polygenic scores and in patients with clinical schizophrenia. The main limitation of the polygenic imaging approach is that it loses the resolution of the individual gene afforded by the single SNP approach. However, this information can be incorporated into the polygenic score method, albeit not at the level of single genes, but at the level of biological pathways (Linden 2012b). For example, the polygenic risk score for schizophrenia can be broken down into genes affecting different neurotransmitter pathways. Genes coding for proteins in the postsynaptic density, which incorporates the NMDA receptor into the cell membrane, are particularly enriched for schizophrenia risk variants. Both functional (including neurophysiology) and structural imaging will have an attractive application in elucidating the effects of this cumulative genetic risk in the pathway for synaptic plasticity on brain and behaviour (Hall et al. 2015).

References

Fleisher AS, Chen K, Quiroz YT, et al. (2012) Florbetapir PET analysis of amyloid- deposition in the presenilin 1 E280A autosomal dominant Alzheimer's disease kindred: a cross-sectional study. *Lancet Neurol*, **11**(12):1057–65.

Hall J, Trent S, Thomas KL, O'Donovan MC, Owen MJ (2015) Genetic Risk for Schizophrenia: Convergence on Synaptic Pathways Involved in Plasticity. *Biol PsychiatryI*, 77(1):52–8.

Jack CR, Knopman DS, Jagust WJ, et al. (2010) Hypothetical model of dynamic biomarkers of the Alzheimer's pathological cascade. *Lancet Neurol*, **9**(1):119–28.

Lancaster TM, Brindley LM, Tansey KE, et al. (2014) Alzheimer's risk variant in CLU is associated with neural inefficiency in healthy individuals. *Alzheimers Dement,* **Pt ii**: S1552–60(14)02867-2.

Linden DE (2012a) *The Biology of Psychological Disorders.* Basingstoke and New York: Palgrave Macmillan.

Linden DE (2012b) The challenges and promise of neuroimaging in psychiatry. *Neuron,* **73**(1):8–22.

Owen MJ (2014) New Approaches to Psychiatric Diagnostic Classification. *Neuron,* **84**(3):564–71.

Poels EM, Kegeles LS, Kantrowitz JT, et al. (2014) Imaging glutamate in schizophrenia: review of findings and implications for drug discovery. *Mol Psychiatry,* **19**(1):20–9.

Prvulovic D, Van de Ven V, Sack AT, Maurer K, Linden DE (2005) Functional activation imaging in aging and dementia. *Psychiatry Res,* **140**(2):97–113.

Reiman EM, Quiroz YT, Fleisher AS, et al. (2012) Brain imaging and fluid biomarker analysis in young adults at genetic risk for autosomal dominant Alzheimer's disease in the presenilin 1 E280A kindred: a case-control study. *Lancet Neurol,* **11**(12):1048–56.

Rive MM, van Rooijen G, Veltman DJ, Phillips ML, Schene AH, Ruhé HG (2013) Neural correlates of dysfunctional emotion regulation in major depressive disorder. A systematic review of neuroimaging studies. *Neurosci Biobehav Rev,* **37**(10 Pt 2):2529–53.

Wang XJ, Krystal JH (2014) Computational Psychiatry. *Neuron,* **84**(3):638–54.

Chapter 8

Neuroimaging and 'mind reading'

Key points

- Neuroimaging can provide maps of neural correlates of psychiatric symptoms, for example, hallucinations or substance craving
- Some pathologies that can lead to delinquent behaviour can be identified by clinical imaging
- However, it is currently not possible to identify personality traits or confirm specific symptoms on the basis of imaging
- Ethical caveats of psychiatric imaging include the potential use of predictive information by insurance companies and the early identification of disease risk in the absence of preventive treatments

8.1 Can neuroimaging read out mental states and psychiatric symptoms?

8.1.1 Symptom mapping

One of the most powerful and evocative imaging approaches to psychiatric disorders is the direct mapping of neural correlates of symptoms, which is mainly the domain of fMRI because this technique has the ideal combination of spatial and temporal resolution. Such a window into the mind can help elucidate the pathophysiological mechanisms behind some of the most disturbing symptoms—and possibly provide new treatment targets (see Chapter 10)—but also give patients insights into the underpinnings of their own mental experience, which they often find helpful. Symptom mapping can be conducted under controlled experimental conditions if it is feasible and ethically acceptable to induce symptoms during scanning, for example, craving induction by presentation of drug-related cues, induction of phobic symptoms, or induction of sadness. In these cases appropriately trained clinical researchers can apply neutralization procedures that prevent long-term effects of the induced symptoms. However, this approach is not advisable for symptoms that can easily spiral out of control, such as hallucinations and other psychotic symptoms. For this group of symptoms, investigators will have to wait for symptoms to occur spontaneously and then make the scanning procedure as comfortable as possible for the patient. In such a setting it is possible to ask patients to report the on- and offsets of hallucinations by releasing a button and recording the corresponding fMRI signal. The time course of reported hallucinations can then be used to model brain activity at the whole-brain level and the resulting correlation maps reveal areas with increased or decreased activity during voice hearing. Areas with increased activity during auditory hallucinations included the primary auditory cortex (see Fig. 8.1), which is also activated by any incoming auditory stimulus, and the human voice area, which responds selectively to the sound of the human voice (Dierks et al. 1999, Jardri et al. 2011, and Linden et al. 2011).

The wider network associated with auditory hallucinations included areas in the frontal lobe responsible for speech production (supporting models according to which verbal hallucinations derive from inner speech), and limbic areas involved in emotion and memory such as the hippocampus. In patients with schizophrenia, hallucinations have mostly been studied in the auditory domain, but preliminary findings from the study of visual hallucinations indicate that these, too, recruit sensory areas (in the visual cortex) (Oertel et al. 2007). Preliminary studies have also suggested that it may be possible to extract the brain activation patterns associated with hallucinations without the need to recur to online self-report, for example, through data-driven analysis techniques like independent components analysis (van de Ven et al. 2005). Further work is needed to validate this approach, but if this became a possibility, we might be able to deduce the presence of auditory hallucinations from the activity in the auditory cortex—of course only in the absence of actual physical auditory stimuli. The limitations of this approach are that the discovery of brain activation patterns associated with hallucinations still relies on corroboration

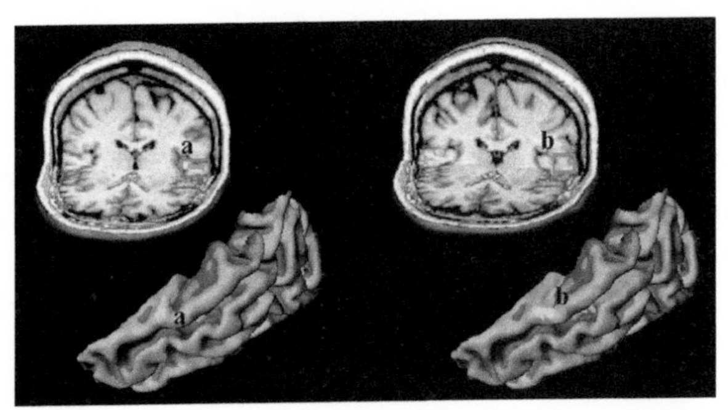

Fig. 8.1 Symptom mapping of auditory hallucinations. These coronal/axial cuts into the brain and surface reconstructions of the temporal lobe show activation in the auditory cortex (Heschl's gyrus) during auditory hallucinations (a) and external auditory stimulation (b) in a patient with chronic schizophrenia. The left side of the brain is on the right.

Adapted from *Neuron*, 22, 3, Dierks T et al, 'Activation of Heschl's gyrus during auditory hallucinations', pp. 615–621. Copyright (1999) with permission from Elsevier.

through patients' own reports and that these patterns may be less specific than needed for the reverse inference from brain activation onto mental experience. For example, auditory imagery activates areas that closely overlap with those active during hallucinations. Furthermore, the brain activation patterns that can be identified with this symptom mapping approach may only—at best—indicate the presence of a symptom, but not whether it distresses the patient. For example, hallucination-related brain activity is very similar in patients with schizophrenia and in non-clinical hallucinators, such as people who report hearing very vivid voices but are not distressed by them and do not seek treatment (Diederen et al. 2011).

8.1.2 **Symptom provocation and cue reactivity**

Symptoms of anxiety disorders can be induced in the laboratory with procedures familiar from exposure therapy. For example, the experimenter might show spiders to patients with spider phobia or boiling kettles to patients with obsessive compulsive disorder (OCD). FMRI and PET studies with this methodology have revealed increased activity in the right caudate and bilateral orbitofrontal cortex (OFC) during the experience of symptoms of OCD, and additionally insula activation when contamination fear is prominent. OFC and insula were also activated during the provocation of phobic symptoms, which is also consistently accompanied by amygdala activation (Linden 2006). Both autobiographical scripts and visual material can be used to induce sadness as an experimental model of some aspects of depression. Several limbic and paralimbic areas, most consistently in the anterior cingulate gyrus and the amygdala, are activated during such states of induced sadness. Based on this approach, the subgenual part of the anterior cingulate gyrus has been identified as a target for deep brain stimulation (DBS) in depression (see Chapter 10).

Addictions are well suited to functional imaging research because states of craving, which are relevant for the understanding of the motivational drive behind the substance abuse, can be induced in the laboratory with a variety of procedures. For example, measuring cue reactivity by showing patients with alcohol use disorders pictures of their favourite drinks can reveal altered patterns of brain activation, even during periods of abstinence. Patients with alcohol dependence and other problem drinkers show heightened activation of areas involved in the processing of reward during such cue reactivity procedures, for example, in the ventral striatum (Wrase et al. 2007, George et al. 2001, and Ihssen et al. 2011). Interestingly, the spatial patterns of hyperactivity overlap across substances of abuse (alcohol, nicotine, cocaine) (Kűhn and Gallinat 2011). Some preliminary evidence also suggests that non-substance-related addictions such as pathological gambling (originally classified as impulse control disorder, but subsumed under the addictive disorders in the DSM-5) are also associated with hyperactivation of the ventral striatum. This is an example where the understanding of the biological circuits underlying particular symptoms and behaviours can support a more rational classification of mental disorders.

8.2 Imaging in a forensic context

8.2.1 Can imaging detect criminal psychopaths?

Each individual has characteristic patterns of thought, beliefs, emotion, and behaviour that are stable across time and can thus be regarded as *traits* (to be distinguished from time-varying *states*, for example, hallucinations). Although some of these traits, for example, impulsivity, may result in directly observable behaviour, most of them are measured by interviews or personality questionnaires. Revealing those 'hidden' traits by neuroimaging would have considerable clinical and forensic implications. One example is the propensity to delinquent behaviour, which is captured by the construct of psychopathy (Hare and Neumann 2008) and can be measured with the revised Psychopathy Checklist (PCL-R) (Hare and Neumann 2009). This psychometric instrument assesses interpersonal and affective attitudes, lifestyle, and antisocial behaviour. Like any measure that is based on questionnaires or, in this case, an interview with a clinician, the outcome depends on the truthfulness of the person tested. Although psychologists have devised ways of detecting lies in such tests psychopaths, in particular, may be good at hiding their real intentions. Identifying from a brain scan whether someone is a psychopath might therefore have important implications for the education and justice systems. The first step towards this would be to identify consistent correlates of psychopathy in the brain. Much of the neuroimaging work in this area has focused on parts of the frontal lobe, the OFC, and the ventromedial prefrontal cortex (VMPFC) because patients with lesions in this region can resemble psychopaths in several dimensions (Gao et al. 2009). Several studies, indeed, found reduced grey matter volume and altered brain metabolism in the prefrontal cortex of individuals with psychopathy. However, patients with lesions to these areas rarely display the levels of ruthlessness found in many psychopaths and, again unlike psychopaths, they are severely impaired in their planning abilities (Kiehl 2006). This makes it difficult to infer a particular type of impaired social functioning, the psychopathic variant, from changes in the structure or function of particular parts of the frontal lobes.

Moreover, there may be simpler ways of objectively ascertaining psychopathy than through brain imaging. Autonomic physiological responses, such as the electrodermal activity (also called skin conductance response, SCR), can be used to measure arousal levels and reactivity to emotional stimuli. One study found less pronounced arousal responses during anticipation of aversive stimuli in psychopathic offenders (Herpertz et al. 2001). Such findings would be compatible with theories that assume reduced fear and sensitivity to punishment in psychopaths, which may make them prone to risky behaviour and less likely to adjust their behaviour based on previous failures or social sanctions (Gao et al. 2009). However, these measures are not stable enough yet to be used as diagnostic tests. Thus, at present, associations between neuroimaging or psychophysiological markers and specific patterns of thought or behaviour are not consistent enough to replace more traditional clinical and personality assessments.

8.2.2 Can imaging support a defence of insanity?

Whether neurobiological changes predictive of criminal behaviour can be extracted from neuroimaging data is a very relevant question for legal practice (Silva 2009) and forensic psychiatry and psychology. A person who committed a violent offence might have the defence of insanity available to them if at the time of the offence they were suffering from a mental illness that precluded them from understanding the nature or wrongfulness of their action or, if they had such understanding, from acting upon it. The first scenario, also called 'cognitive insanity', may occur, for example, in patients with dementia or delusions. In such cases the forensic psychiatrist would have to establish a diagnosis of a recognized mental (e.g. schizophrenia) or neurological (e.g. brain tumour) disease and then show that this disease led to a functional impairment resulting in cognitive insanity (see also Chapter 5 on the diagnosis of parasomnias and epileptic states). The latter scenario, also called 'volitional insanity', is more controversial and less widely used: it would apply in severe cases of impulse control disorders. Objective diagnostic criteria for such disorders obtained through neuroimaging would aid the process of identifying those offenders whose moral judgements and actions were impaired owing to a pathological process in their brain. However, as explained above, such criteria will be maximally as good as the conventional diagnoses on which they are based. In neurobiological terms, it may be more meaningful to use brain imaging to try and predict more specific traits, such as tendency to impulsive behaviour, risk taking, or disregard for the feelings of others.

8.3 Ethics of brain reading

Beyond the potential forensic use, psychiatric brain reading may have wider implications in general clinical settings. One question will be who would benefit from this type of information in clinical practice? The beneficiaries could be the patients themselves if an accurate prediction of disease risk would lead to the implementation of preventive treatment (which is currently not available for most psychiatric disorders). Another benefit to patients and their families of such early knowledge about impending illness might be that they could make adjustments to the patient's social and economic situation in advance of the functional impairment. Most people would find predictions of individual disease risk for such purposes morally acceptable, although many might still prefer not to know about such risks. Doctors and biomedical researchers might also benefit if they could predict the individual risk of developing dementia or schizophrenia better because this would allow them to test drugs and other interventions that might in the future prevent these diseases. Of course, by the same token, pharmaceutical companies might also benefit from such information. As soon as such commercial interests get involved people may become more wary about sharing information about their own disease risk. Finally, when it comes to insurance companies requesting information about disease risk, the beneficiaries would be these companies (or their other customers) rather than the person concerned. Although these considerations are still largely theoretical for most neuropsychiatric disorders, they may soon become very practical issues, especially in the context of the current debate about the appropriateness of predictive testing that utilizes statistical genetic information (Zettler et al. 2014).

References

Diederen KM, Daalman K, de Weijer AD, et al. (2011) Auditory Hallucinations Elicit Similar Brain Activation in Psychotic and Nonpsychotic Individuals. *Schizophr Bull*, **38**(5):1074–82.

Dierks T, Linden D, Jandl M, et al. (1999) Activation of Heschl's gyrus during auditory hallucinations. *Neuron*, **22**(3):615–21.

Gao Y, Glenn A, Schug R, Yang Y, Raine A (2009) The neurobiology of psychopathy: a neurodevelopmental perspective. *Can J Psychiatry*, **54**(12):813–23.

George MS, Anton RF, Bloomer C, et al. (2001) Activation of prefrontal cortex and anterior thalamus in alcoholic subjects on exposure to alcohol-specific cues. *Arch Gen Psychiatry*, **58**(4):345–52.

Hare RD, Neumann CS (2008) Psychopathy as a clinical and empirical construct. *Annu Rev Clin Psychol*, **4**:217–46.

Hare RD, Neumann CS (2009) Psychopathy: assessment and forensic implications. *Can J Psychiatry*, **54**(12):791–802.

Herpertz S, Werth U, Lukas G, et al. (2001) Emotion in criminal offenders with psychopathy and borderline personality disorder. *Arch Gen Psychiatry*, **58**(8):737–45.

Ihssen N, Cox WM, Wiggett A, Fadardi JS, Linden DE (2011) Differentiating heavy from light drinkers by neural responses to visual alcohol cues and other motivational stimuli. *Cereb Cortex*, **21**(6):1408–15.

Jardri R, Pouchet A, Pins D, Thomas P (2011) Cortical activations during auditory verbal hallucinations in schizophrenia: a coordinate-based meta-analysis. *Am J Psychiatry*, **168**(1):73–81.

Kiehl K (2006) A cognitive neuroscience perspective on psychopathy: evidence for paralimbic system dysfunction. *Psychiatry Res*, **142**(2–3):107–28.

Kühn S, Gallinat J (2011) Common biology of craving across legal and illegal drugs—a quantitative meta-analysis of cue-reactivity brain response. *Eur J Neurosci*, **33**(7):1318–26.

Linden DE (2006) How psychotherapy changes the brain—the contribution of functional neuroimaging. *Mol Psychiatry*, **11**(6):528–38.

Linden DE, Thornton K, Kuswanto CN, Johnston SJ, van de Ven V, Jackson MC (2011) The brain's voices: comparing nonclinical auditory hallucinations and imagery. *Cereb Cortex*, **21**(2):330–7.

Oertel V, Rotarska-Jagiela A, van de Ven VG, Haenschel C, Maurer K, Linden DE (2007) Visual hallucinations in schizophrenia investigated with functional magnetic resonance imaging. *Psychiatry ResI*, **156**(3):269–73.

Silva JA (2009) Forensic psychiatry, neuroscience, and the law. *J Am Acad Psychiatry Law*, **37**(4):489–502.

van de Ven VG, Formisano E, Röder CH, et al. (2005) The spatiotemporal pattern of auditory cortical responses during verbal hallucinations. *Neuroimage* **27**(3):644–55.

Wrase J, Schlagenhauf F, Kienast T, et al. (2007) Dysfunction of reward processing correlates with alcohol craving in detoxified alcoholics. *Neuroimage*, **35**(2):787–94.

Zettler PJ, Sherkow JS, Greely HT. 23andMe, the Food and Drug Administration, and the future of genetic testing. *JAMA Intern Med.* 2014;174(4):493–494.

Chapter 9

Imaging of treatment effects

Key points

- Radionuclide imaging can assess whether a drug engages with its target in the brain and help determine dose-response relationships
- Neuroimaging can be used to search indicators of treatment response ('surrogate markers') or markers that predict treatment success
- Functional neuroimaging has revealed changes in brain networks after psycho- and pharmacotherapy and also after invasive (deep brain stimulation, psychiatric surgery) treatments

There is an obvious place for imaging in many areas of surgical and medical intervention. A surgeon may want to confirm the successful treatment of a fractured bone with a follow-up X-ray, or an oncologist may want to order an FDG-PET scan to check whether a chemotherapy of metastatic cancer has been successful. Because psychiatric treatments do not directly target identifiable organic pathology, imaging presently has no such role in clinical treatment validation in psychiatry. However, research into the use of neuroimaging for the detection and monitoring of treatment effects has greatly expanded over the last decade, and first studies have explored whether imaging parameters can predict who will respond to what type of treatment and thus aid patient stratification. Finally, neuroimaging already has a place in the design and monitoring of brain modulation treatments, which will be discussed Chapter 10.

9.1 Radionuclide imaging and pharmacotherapy

Most psychiatric drugs target membrane proteins (receptors or transporters). Before the introduction of a new drug a manufacturer generally has to demonstrate its mechanism by pre-clinical pharmacology studies, both in vitro and in vivo (in animals). However, this does not routinely include in vivo studies in humans that show that the drug actually engages with the target molecule, and at what doses occupancy of the receptor or transporter is achieved most effectively. Yet the demonstration of such 'target engagement' with SPECT or PET can be clinically very important, especially when there is a mismatch between the effects in animal models and in patients.

The radioligand ^{11}C-raclopride is a PET tracer that binds to D2 and D3 dopamine receptors. It can be used to assess target engagement of drugs that are supposed to block these receptors, for example, antipsychotic drugs. Receptor occupancy by these drugs can be calculated from the reduction in the PET signal in the same person after the administration of the drug, or by comparing the uptake of the tracer between a group of treated individuals and untreated control participants (see Chapter 2, Sections 2.3 and 2.4.4). PET imaging with raclopride has provided the basis for the understanding of the link between D2 receptor occupancy and therapeutic effects (and side effects) of typical antipsychotic drugs. For most drugs, the therapeutic window is in the range of 60–80% D2 receptor occupancy in the striatum, and levels above 80% are associated with extrapyramidal side effects. Less information is available about atypical antipsychotics, but as expected from its pharmacology, clozapine in therapeutic doses is associated with much lower D2/D3 receptor occupancy (Nord and Farde 2011). Target engagement has also been demonstrated with PET or SPECT for most antidepressants. The main ligand for this application is ^{11}C-labelled 3-amino-4-(2-diethylaminomethylphenylsulfanyl)-benzonitrile (DASB), which binds to the serotonin transporter (abbreviated to 5-HTT or SERT). For example, the SERT occupancy during treatment with therapeutic doses of selective serotonin reuptake inhibitors is generally in range between 70% and 85%. Such findings are clinically relevant because they help determine the optimal dose range for an existing drug, although ultimately this will be decided by the profile of clinical effects and side effects. Target engagement studies can also be used to demonstrate or reject putative mechanisms of action of psychotropic drugs (see also Chapter 6, Nutt and Nestor 2013). For example, one

PET study confirmed blockade of the monoamine oxidase A enzyme (MAO-A) after treatment with moclobemide, confirming its role as monoamine oxidase (MAO) inhibitor, but did not find a similar mechanism for treatment with St John's wort, which had been presumed to be an MAO inhibitor as well (Sacher et al. 2011). This use of neuroimaging to demonstrate target engagement is not limited to therapeutic agents but has also been extended to the pharmacokinetic profile of drugs of abuse (see Chapter 5, Nutt and Nestor 2013).

One important limitation of the target engagement approach is that not all antidepressants meet the expected profile of in vivo pharmacology. Occupancy levels can be higher or lower than expected. For example, the tricyclic antidepressant clomipramine, a relatively pure serotonin reuptake inhibitor, can occupy 80% of the SERT even at sub-therapeutic doses. Conversely, therapeutic levels of atomoxetine, a selective noradrenaline reuptake inhibitor (SNRI) and bupropion, a dopamine reuptake inhibitor (DRI) were accompanied by only low levels of noradrenaline (NET) or dopamine transporter (DAT) occupancy (Wong et al. 2009). Such findings are particularly relevant for the development of new drugs because they suggest that established drugs with a supposedly known mechanism of action may actually be operating through as yet undiscovered molecular mechanisms.

The development of new drugs in psychiatry faces two major difficulties that could potentially be alleviated with the help of neuroimaging. Firstly, the dearth of diagnostic and mechanistic biomarkers (see Chapter 6) makes rational drug development very difficult. Most compounds that were introduced to the market for mental disorders in the last 20 years were variations on older drugs. Secondly, the lack of biomarkers for treatment effects, so called surrogate markers, make clinical trials lengthy and costly. These difficulties have led several major pharmaceutical companies to withdraw from the development of new psychotropic drugs over the last years.

Imaging can aid psychiatric drug development in several ways. If a mechanistic biomarker has been identified it can help the rational design of new compounds that target this specific mechanism. Once pre-clinical evidence has been accumulated to take this compound into clinical testing, neuroimaging can be employed at a very early state (even at phase 0) to demonstrate that the substance has the desired pharmacodynamic and pharmacokinetic properties, for example, that it passes through the blood brain barrier. Although, as discussed in Chapters 6 and 7, no universally accepted mechanistic biomarkers have been established for any of the major mental disorders, recent research based on animal models has been pursuing new pharmacological avenues, for example, the modulation of metabotropic glutamate receptors for cognitive and negative symptoms of schizophrenia. Radiotracers for several metabotropic glutamate receptors are currently in development. Once they are available for routine use in humans they could be employed to probe any putative over- or underexpression of metabotropic glutamate receptors in patients with schizophrenia and demonstrate that new drugs developed on the basis of this pharmacological model actually interact with the targeted receptors.

Imaging can also be used to determine the dosages to be tested in phase 2 and 3 trials (see Table 9.1 on the definition of trial phases). If a drug has an established therapeutic mechanism, its dosage can be benchmarked against clinically effective target occupancy of established drugs. For example, a company trying to bring a new antipsychotic drug to market that is supposed to be clinically as effective as existing typical antipsychotics but purported to have fewer side effects could test a dose range that results in 60–80% occupancy of D2 receptors, similar to that achieved by established substances. If new drugs fail in phase 2 trials there is always the concern that the dose may not have been sufficiently high. However, if a drug fails to show a clinical effect in phase 2, although it had sufficient target engagement, it can be assumed that the dose was probably high enough but that the pharmacological target was not relevant to the disease, and the testing of this drug can be abandoned without further dose adjustment, and thus without further expensive and probably futile clinical trials (Wong et al. 2009). Finally, imaging can provide surrogate markers of the treatment's effects—where a change in an imaging parameter predicts a clinical response (see Table 9.2). Such surrogate markers can serve as endpoints for clinical trials, and massively reduce their cost and duration, both by establishing and rejecting effectiveness. The classical example of a neuroimaging marker that has been proposed as an endpoint for clinical trials is lesion load in multiple sclerosis, measured by

Table 9.1 Phases of clinical trials

Phase	Description
0	Exploratory study to determine pharmacodynamic and pharmacokinetic properties of a new drug with small, subtherapeutic doses in a small number (usually around ten) of volunteers
1	Safety study in approx. 20–100 healthy volunteers to determine most common adverse events and find the dose range for later phases
2	Study in up to 300 patients that gathers preliminary data on effectiveness of therapeutic doses of the substance/ intervention; can be conducted as randomized controlled trial (RCT) with placebo arm. This phase also continues to assess safety.
3	Study (generally RCT) that aims to provide definitive data on effectiveness by comparison with placebo intervention or other established treatments. The new therapy is often also studied in combination with other therapies. This phase also continues to assess safety. Several thousands of patients can be included.
4	Post-approval study that gathers additional information about the safety, efficacy, or optimal use of the new therapy.

MRI. Although no such marker is in sight for any of the mental disorders, there is very active research interest in PET surrogate markers of treatment success in AD. First results on amyloid reduction, as measured with the amyloid-binding ^{11}C-Pittsburgh compound B (PiB), by several experimental treatments for AD were promising, but recent trials of the monoclonal antibody bapineuzumab produced conflicting results (Salloway et al. 2014). Although the treatment halted the progression of amyloid accumulation in carriers of the ApoE epsilon 4 risk variant, it did not result in any clinical improvement. Such a finding is still very relevant for the earlier stages of drug development because, as explained above, repeated failure to see clinical effects in the presence of the desired biological effect would call the presumed therapeutic mechanism into doubt. Specifically, if further trials were to demonstrate successful halting or even reduction of amyloid accumulation without corresponding clinical effects, the strategy of developing immune therapy targeting amyloid plaques in patients with already manifest AD would have to be reconsidered.

Table 9.2 Biomarker types and examples from neuroimaging

Type of biomarker	Criteria	Applications	Successful examples	Currently investigated candidates (examples)
Diagnostic	Reliable difference between cases and general population (or between different disease groups)	Aid diagnosis	FDG-PET in AD	Multivariate analysis of structural imaging for diagnosis of autism (Ecker 2011)
Prognostic	Level of biomarker is predictive of course of illness or treatment response	Aid prognosis; aid treatment decisions		Multivariate analysis of imaging parameters to predict conversion from prodromal schizophrenia (Koutsouleris et al. 2014); fMRI (emotion tasks and at rest) and prediction of antidepressant response (McGrath et al. 2014)
Proof of concept/ mechanism	Confirms the putative mechanism of action of a drug	Dose finding for clinical trials; therapeutic drug monitoring	11C-raclopride PET of D2/D3 receptor occupancy by antipsychotic drugs	PET ligands of metabotropic glutamate receptors?
Surrogate	Change in biomarker is predictive of clinical outcome	Use as endpoints in clinical trials (to make trials faster and cheaper)	White matter lesion load in MS (currently still as secondary outcome measure, in addition to clinical measures)	^{11}C- or ^{18}F-labelled ligands for amyloid for intervention trials in AD

Source data from: Wong et al. 2009, copyright (2009) Nature Publishing Group; Lesko LJ and Atkinson AJ (2001), copyright (1995) Annual Reviews.

9.2 Therapeutic drug monitoring

Radionuclide imaging can also have a useful role once a drug has been clinically evaluated and introduced. Interindividual differences in the clinical response can vary greatly. One reason for non-response or adverse effects may be under- or overdosing. Although this can be monitored to some degree through plasma concentrations of the drug, in vivo molecular imaging is needed to determine the relationship between plasma concentrations and target occupancy in the brain and can help refine the recommended range of plasma levels (obviously within the constraints of clinical safety). Although it is much cheaper and easier to measure plasma concentrations of a drug than the target occupancy in the brain, radionuclide imaging may still be helpful in individual cases because of individual differences in pharmacodynamics and pharmacokinetics. For example, variability in the expression of neurotransmitter receptors or transporters in the brain may result in differences in the clinical responses to the same dose of a drug across individuals. However, in clinical practice the mainstay of therapeutic drug monitoring is still the measurement of plasma concentrations, which may allow broad inferences on target occupancy in the brain, rather than individual radionuclide imaging. Another potential source of discrepancies between plasma concentrations and central action are interindividual differences in the blood brain barrier. For example, there is preliminary evidence that the function of p-glycoprotein (p-gp), one of the main pump proteins of the blood brain barrier, is impaired in aging and neurodegenerative diseases. This, in turn, can be measured with PET, using ^{11}C-labelled verapamil, which binds to p-gp. Again, this is only a research application at present, and no clinical tests to quantify blood brain barrier function to help with dose findings are available. However, with increasing interest in personalized medicine and pharmacogenetics, the use of imaging (and other biomarkers) to determine individualized dose regimes will remain a topic of active research. (For the related area of imaging research into the pharmacological and psychological effects of substances of abuse, see Chapter 7, Nutt and Nestor 2013.)

9.3 Therapy effects on functional and structural imaging measures

The stress model of depression predicts that chronic hyperactivation of the Hypothalamic-Pituitary-Adrenal (HPA) axis in depression damages neurons in the hippocampus. Volume loss in the hippocampus has indeed been observed in patients with depression, especially those with a longstanding illness, although the findings of structural imaging studies have not been consistent. This model can be extended to predict that antidepressants, possibly through the activation of neurotrophic factors, promote neuroplasticity and reverse this structural damage. Evidence for this stems from rodent studies and some human studies comparing hippocampal volume before and after treatment (Höflich 2012).

The modulation of neurotransmitter metabolism and activity that is attained through psychotropic drugs might also be reflected in changes in brain activity and metabolism downstream from the immediate neurochemical effects. Changes in blood flow and oxygenation can be measured with PET using ^{15}O-labelled water and with fMRI, and changes in glucose metabolism can be measured with FDG-PET. One instance where PET studies have indicated both a specific abnormality before treatment and a treatment-related normalization is the reduction of hypermetabolism in the subgenual cingulate gyrus after antidepressant treatment (Mayberg et al. 1999), and a similar pattern of pre-treatment hyperactivity in this area that was reduced by therapy has recently been described for a 16-week psychodynamic psychotherapy (McGrath et al. 2013). However, this approach can produce meaningful results even if the imaging method employed does not bring up any abnormality before treatment because it is likely that psychiatric treatments work, at least partly, through the activation of compensatory mechanisms. Thus a particular imaging modality may not be sensitive to the abnormalities in brain activation contributing to a state of mental illness, but to the compensatory processes induced by medication and other treatments.

9.4 Neuroimaging and psychotherapy

The idea that psychiatric treatment may activate compensatory processes or re-balance the brain rather than directly correcting a neurochemical deficit is very relevant to the evaluation of non-pharmacological treatments. The investigation of the neural changes induced by different forms of psychotherapy is of great importance for the understanding of their mechanism of action. This can help refine therapeutic protocols and make them more efficient, for example, by focusing on the cognitive processes that have the largest effect in terms of desirable outcomes in brain network activation (assuming that these are known). Knowing the neural (including neurochemical) effects of psychotherapy would also facilitate its rational combination with drugs. It is very common practice, particularly in the treatment of depression, to combine psychological with pharmacological interventions at the same time, but very little is known about how these two approaches might complement each other. Studies that directly compare brain effects of psychotherapy with those of psychiatric drugs are an important first step in this direction.

The treatment of social phobia lends itself particularly well to such comparisons because instant effects can be obtained with high success rates with either behavioural therapy or drug treatment. In a PET study of social phobia, a similar reduction in limbic activity during preparation of a public speech was observed in patients who were treated with antidepressants or a behavioural intervention. Converging effects of behavioural therapy and SSRIs on brain activation were also observed in OCD. Both treatments led to a reduction of previously increased resting blood flow in the basal ganglia. These results are conceptually very interesting because they suggest a specific chemical mechanism—serotonergic activation—for a behavioural intervention. However, this would need to be corroborated by evidence from direct molecular techniques, e.g. PET, because both treatments may also converge downstream from the serotonergic effects (Linden 2006). Effects of psychotherapy and pharmacotherapy have also been compared in patients with depression. The number of studies has so far been too low to allow for definitive conclusions, but both convergent and divergent findings have been reported. Venlafaxine and other antidepressants reduce glucose metabolism (and thus presumably neural activity) in the posterior part of the subgenual cingulate gyrus fairly consistently across studies, whereas CBT had the opposite effect in an adjacent area (Kennedy et al. 2007). At the level of patients' (at least initial) experience psycho- and pharmacotherapy are very different. Psychotherapy utilizes concepts of emotion control and self-efficacy to a much larger extent than pharmacotherapy. One would thus expect some divergence in the neuro-psychological processes underlying each technique's effect.

9.5 **Brain imaging of brain stimulation and psychiatric surgery**

Electroconvulsive treatment (ECT) is the most effective antidepressant treatment. It operates through the induction of a generalized seizure but exactly how and why this might be beneficial is unknown. In fact, frequent seizure activity in chronic epilepsy can lead to hippocampal damage, which is not desirable in depression and can produce memory side effects. Based on animal models it has long been speculated that ECT promotes neurotransmitter release, and radionuclide studies have indeed produced some evidence for activation of serotonergic and dopaminergic neurotransmission during and after ECT (Baldinger et al. 2014). One of the main limitations of ECT is that its clinical effects are rather short-lived. One use for imaging in the development of transfer protocols (for example, using tDCS) that enable ECT effects to last longer would be in the identification of analogous neurochemical changes.

Little is known about the effects of psychiatric deep brain stimulation or stereotactic lesion surgery beyond the immediately affected area. Both approaches are being used as experimental treatments for severe, otherwise intractable forms of depression and OCD (Linden 2014). One FDG-PET study directly compared the metabolic effects of DBS of the bed nucleus of the stria terminalis and capsulotomy (lesioning the anterior limb of the internal capsule that carries fibres from the thalamus to frontal cortex but not the pyramidal tract) for OCD (Suetens et al. 2014). This study found evidence for a partial functional disconnection between the frontal lobe and the thalamus after capsulotomy.

9.6 **Neuroimaging and treatment response prediction**

Although most people with mental disorder respond to some form of pharmacotherapy, only a minority of patients respond to the first treatment. It is very usual for patients to try several antidepressants or antipsychotics before they find the substance or combination of substances that suits them. It is very difficult to predict from clinical parameters alone whether a patient will respond better to a primarily serotonergic or noradrenergic antidepressant, for example. Predictive biomarkers of treatment response would thus be hugely beneficial in reducing the cost and suffering associated with failed treatment attempts. Although there is indeed some tentative evidence from PET studies that higher SERT availability is associated with better response to selective serotonin reuptake inhibitors (SSRI) the effects are not strong and stable enough to support clinical differential treatment decisions. Several fMRI and PET studies also found that higher activity in the rostral anterior cingulate cortex (ACC) before treatment, either at rest or in response to emotional tasks, predicted better treatment outcomes after a variety of antidepressants (Höflich et al. 2012).

Although such a general prediction of treatment effects would be interesting because it points to potential shared therapeutic mechanisms it is not particularly useful clinically, where the main question in manifest depression is generally not whether to treat but how. The first studies have started to investigate the potential use of imaging markers for such treatment stratification in psychiatry. In one trial comparing escitalopram and CBT, patients with lower relative glucose metabolism in the insula had better responses to CBT and those with higher insula metabolism responded better to the drug (McGrath et al. 2013). However, the predictive accuracy of these metabolic patterns needs to be verified in independent samples because it is always much easier to find a discriminating feature in a given sample than to replicate it in a new group of patients, which is the central test for clinical utility of a predictive marker.

9.7 Neuroimaging and side effects

Side effects of drugs are a major obstacle to the management of psychiatric patients and severely reduce compliance. Their mechanisms are generally poorly understood, which makes it hard to design safer alternatives. The extrapyramidal side effects of antipsychotic drugs are related to high occupancy levels of dopamine receptors (see Section 9.1), and the sedative effects of many psychiatric drugs are generally attributable to anticholinergic and antihistaminergic effects. One example of a common side effect with unknown mechanism is sexual dysfunction on treatment with antidepressants, particularly SSRIs. Patients who had developed sexual dysfunction on treatment with the SSRI paroxetine showed lower activation of areas in the midbrain and frontal lobe when exposed to erotic video clips than patients on placebo or bupropion. Although fMRI lacks neurochemical resolution these findings may provide indirect evidence for a dopamine-related mechanism for SSRI-related sexual dysfunction because a similar attenuation of reward responses has been observed after administration of antidopaminergic drugs (Graf et al. 2014). The potential use of imaging techniques for the investigation of side effects of psychotropic drugs has not been fully explored yet and this research area entails attractive opportunities for future work.

References

Baldinger P, Lotan A, Frey R, Kasper S, Lerer B, Lanzenberger R (2014) Neurotransmitters and electroconvulsive therapy. *J ECT*, **30**(2):116–21.

Ecker C (2011) Autism biomarkers for more efficacious diagnosis. *Biomark Med*, **5**(2):193–5.

Graf H, Walter M, Metzger CD, Abler B. Antidepressant-related sexual dysfunction—perspectives from neuroimaging. *Pharmacol Biochem Behav*, **121**:138–45.

Höflich A, Baldinger P, Savli M, Lanzenberger R, Kasper S (2012) Imaging treatment effects in depression. *Rev Neurosci*, **23**(3):227–52.

Kennedy S, Konarski J, Segal Z, et al. (2007) Differences in brain glucose metabolism between responders to CBT and venlafaxine in a 16-week randomized controlled trial. *Am J Psychiatry*, **164**(5):778–88.

Koutsouleris N, Riecher-Rössler A, Meisenzahl EM, et al. (2014) Detecting the Psychosis Prodrome Across High-risk Populations Using Neuroanatomical Biomarkers. *Schizophr Bull*, **10**(1093).

Lesko LJ, Atkinson AJ (2001) Use of biomarkers and surrogate endpoints in drug development and regulatory decision making: criteria, validation, strategies. *Annu Rev Pharmacol Toxicol*, **41**:347–66.

Linden DE (2006) How psychotherapy changes the brain—the contribution of functional neuroimaging. *Mol Psychiatry*, **11**(6):528–38.

Linden DE (2014) *Brain Control*. Basingstoke: Palgrave Macmillan.

Mayberg HS, Liotti M, Brannan SK, et al. (1999) Reciprocal limbic-cortical function and negative mood: converging PET findings in depression and normal sadness. *Am J Psychiatry*, **156**(5):675–82.

McGrath CL, Kelley ME, Dunlop BW, Holtzheimer PE, Craighead WE, Mayberg HS (2014) Pretreatment brain States identify likely nonresponse to standard treatments for depression. *Biol Psychiatry*, **76**(7):527–35.

McGrath CL, Kelley ME, Holtzheimer PE, et al. (2013) Toward a neuroimaging treatment selection biomarker for major depressive disorder. *JAMA Psychiatry*, **70**(8):821–9.

Nord M, Farde L (2011) Antipsychotic occupancy of dopamine receptors in schizophrenia. *CNS Neurosci TherI*,**17**(2):97–103.

Nutt D and Nestor L (2013) *Addiction*, Oxford Psychiatry Library Series. Oxford: Oxford University Press.

Sacher J, Houle S, Parkes J, et al. (2011) Monoamine oxidase A inhibitor occupancy during treatment of major depressive episodes with moclobemide or St. John's wort: an [11C]-harmine PET study. *J Psychiatry Neurosci*, **36**(6):375–82.

Salloway S, Sperling R, Fox NC, et al. (2014) Two phase 3 trials of bapineuzumab in mild-to-moderate Alzheimer's disease. *N Engl J Med*, **370**(4):322–33.

Suetens K, Nuttin B, Gabriëls L, Van Laere K (2014) Differences in Metabolic Network Modulation Between Capsulotomy and Deep-Brain Stimulation for Refractory Obsessive-Compulsive Disorder. *J Nucl Med*, **55**(6):951–9.

Wong DF, Tauscher J, Gründer G (2009) The role of imaging in proof of concept for CNS drug discovery and development. *Neuropsychopharmacology*, **34**(1):187–203.

Chapter 10

Neurophysiological treatments

Key points

- Repetitive transcranial magnetic stimulation (rTMS) is an effective adjunctive treatment for depression, but little is known about its long-term efficacy
- Neuroimaging findings have guided the development of both invasive and non-invasive brain stimulation protocols, for example, rTMS over the temporal lobe for hallucinations
- Patients can train to self-regulate patterns of their own brain activity through real-time feedback of EEG or fMRI signals

10.1 Brain stimulation in mental disorders

This chapter will first review the brain stimulation techniques that have recently attracted great interest (although still relatively little clinical use) within psychiatry (Section 10.1). Their development is closely intertwined with progress in psychiatric imaging. Although only invasive approaches (deep brain stimulation) come with a requirement for neuroimaging, non-invasive transcranial stimulation has also been heavily influenced by neuropsychological models that were based on imaging results. The next section will therefore provide a few examples of brain stimulation protocols that have been informed by knowledge derived from imaging (Section 10.2). Finally, this chapter will discuss ways of directly incorporating imaging or EEG in therapeutic brain modulation protocols through neurofeedback (Section 10.3). Another neurophysiological treatment, ECT, has already been introduced in Chapter 5 (Section 5.4).

10.1.1 TMS in depression

Trials of TMS in depression have been conducted for over 20 years, and by now several thousands of patients have been included. Recent meta-analyses point to clinical effects (improvement in depression rating scales) that are superior to those of placebo treatment (mostly sham TMS) (Berlim et al. 2013), but a major limitation is that little information is available about the sustainability of the effects. After a course of several weeks of TMS is completed, patients generally still take antidepressant medication. Although adjunctive TMS may have lasting benefits in relapse prevention (Janicak et al. 2010) its use as a standalone treatment has not been formally evaluated.

The most commonly used protocols involve high frequency (HF) (10-20Hz) repetitive TMS (rTMS) over the left or low frequency (LF) (1Hz) rTMS over the right frontal lobe. These protocols were based on early functional imaging findings, which indicated hypoactivity of the left and/or hyperactivity of the right frontal lobe in patients with depression, and a normalization of this asymmetry with drug treatment (Martinot et al. 1990). HF rTMS is supposed to activate the targeted brain areas, whereas LF rTMS is supposed to inhibit them, and imaging has been used to confirm this frequency-dependent difference of TMS effects in patients with depression; regional cerebral blood flow, as measured by PET, indeed increased after 20Hz TMS and decreased after 1Hz TMS. However, these effects were not confined to the stimulated hemisphere; ten daily treatments with 1Hz TMS over the left prefontal cortex led to blood flow reductions in the right, rather than left, prefrontal cortex (Speer et al. 2000). Such distant effects of TMS underline the importance of monitoring physiological treatment effects with (concurrent or interleaved) imaging. It is also worth considering that the model of frontal left < right asymmetry is by no means supported by all of the relevant imaging or EEG studies (Linden 2014b). Imaging could therefore have a crucial role in identifying patients who may respond to different types of TMS protocols and establishing the mechanisms underlying treatment effects.

10.1.2 TMS in schizophrenia

Inhibitory (1 Hz) TMS over the left or bilateral temporo-parietal cortex has been used in several studies to treat auditory hallucinations. These interventions, generally at an intensity of 90% of the resting motor threshold (RMT), have lasted for up to 20 sessions and employed up to 1,200 pulses per session. Study results were inconsistent (and any improvements generally shortlived) but still promising enough for a recent consensus statement to conclude that this approach has 'possible efficacy' (Lefaucheur et al. 2014). The other main field of experimental application of TMS in schizophrenia concerns the treatment of negative symptoms. Here, an activating stimulation (generally 10Hz) of left or bilateral prefrontal is chosen, based on the hypofrontality model of negative and cognitive symptoms of schizophrenia. Stimulation intensity in published studies varied between 90% and 110% of RMT. Improvement of negative symptoms was consistent enough for the consensus group to conclude 'probable efficacy', but it was noted that concomitant antidepressant effects of the TMS may contribute to this improvement (Lefaucheur et al. 2014).

10.1.3 **Transcranial electrical stimulation in depression**

Transcranial magnetic stimulation is not the only non-invasive way of inducing electrical field changes in the brain. Currents through the brain can also be induced by placing electrodes on the scalp. In the simplest form one electrode is attached to the anodal pole of a battery and the other to the cathodal pole, and they pass a constant low-intensity current through the brain (transcranial direct current stimulation (tDCS)). It is also possible to change the polarity of the stimulation, resulting in 'transcranial alternating current stimulation' (tACS), which allows probing the effect of specific oscillatory frequencies on brain activity and function. Most psychiatric applications have used tDCS. It is generally assumed that areas under the anode will be depolarized and thus activated, whereas areas under the cathode will be hyperpolarized and thus inhibited (see Section 3.4, Chapter 3). Whereas TMS at commonly used intensities can evoke motor potentials or produce muscle twitches (which can be used to determine the 'motor threshold' for standardization of stimulation protocols), tDCS has no overt effects on motor behaviour because the induced electrical fields are much smaller. It thus relies on neuroimaging for identification of its targets even more than TMS (Nitsche et al. 2008).

Most tDCS protocols for depression employ anodal stimulation over the left prefrontal cortex, based on the theory of left frontal hypoactivation in depression mentioned previously. It is also possible to place the anode over the left prefrontal cortex and the cathode over the right in order to try and rebalance activity across both hemispheres. Such tDCS protocols, generally conducted in daily sessions of about ten minutes over several days or weeks, have shown some effects in depression (and also other psychiatric conditions, for example, several types of addiction) (Kuo et al. 2014). However, the evidence base is much smaller than that for TMS, and similar to TMS further information about the long-term effects and about the mechanisms of action is needed. Neuroimaging will be useful both to determine any dysfunctional patterns of brain activation that can be targeted with TMS or tDCS (in order to target the right areas in the right patients) and to demonstrate the brain changes associated with clinical effects.

10.1.4 **DBS in depression and OCD**

Deep brain stimulation (DBS) is an invasive technique for neuromodulation that involves insertion of electrodes into deep cortical or subcortical areas (another invasive brain stimulation technique is cortical surface stimulation, which has some applications in neurology, particularly in the experimental treatment of refractory pain (Linden 2014a)). The main clinical application of DBS is in Parkinson's disease, where over 60,000 patients have so far been treated with chronic stimulation of the subthalamic nucleus and other targets in the basal ganglia. The psychiatric application of DBS is still experimental, but research programmes in treatment-refractory depression and OCD have included several hundred patients. The development of DBS for depression has been strongly influenced by neuroimaging research. The first target area for DBS in depression was the 'subgenual' part of the cingulate cortex, where it is nested below the knee (Latin: *genu*) of the corpus callosum (Mayberg et al. 2005). Other targets of recent DBS studies in depression were the ventral internal capsule and the nucleus accumbens. Psychiatric DBS protocols generally employ bilateral chronic stimulation and, like those for movement disorders and pain, high stimulation frequencies (over 100 Hz) that are supposed to attenuate activity in the targeted area. In studies reported so far the number of responders did not depend substantially on the stimulated area and was generally between 30% and 50% (Bewernick et al. 2012, Holtzheimer et al. 2012, and Lozano et al. 2008), but this may include a substantial placebo effect because it is difficult to conduct sham intervention studies for invasive procedures (Linden 2014a).

DBS of various parts of the basal ganglia, for example, the nucleus accumbens, subthalamic nucleus or the ventral capsule, has also improved OCD symptoms in some patients (Greenberg 2010). One way of addressing the contribution of the placebo effect is to implant electrodes and then conduct comparisons between periods when they are switched on and when they are switched off. In one such 'cross-over' trial with 16 OCD patients, who had electrodes implanted into the subthalamic nucleus bilaterally, symptom scores during the stimulation phases were indeed significantly lower than during the 'off' phases, but still amounted to OCD of moderate severity (Mallet et al. 2008). Furthermore, about one third of OCD of patients in DBS studies suffered serious surgical or psychological adverse events (Kisely et al. 2014).

10.2 How imaging informs new therapies

The use of neuroimaging for treatment may go beyond the monitoring and prediction of clinical effects discussed in Chapter 9. Functional imaging has already informed the selection of stimulation sites for both invasive and non-invasive interventions (see Section 10.1). For example, the TMS protocols targeting the temporal lobe for auditory hallucinations are based on fMRI and ASL studies implicating the auditory cortex in the generation of hallucinations. The basis to this is that if overactivity of a brain region contributes to a symptom, an attenuation of its activity through inhibitory brain modulation could improve this symptom. The development of deep brain stimulation (DBS) for depression was guided by a similar idea. The first target, the white matter adjacent to the subgenual cingulate cortex, was selected because of PET findings of increased blood flow and glucose metabolism in this area during episodes of depression (Mayberg et al. 2005). Later DBS studies in depression added two subcortical regions: the nucleus accumbens and the ventral capsule, to the array of potential targets (see Fig. 10.1).

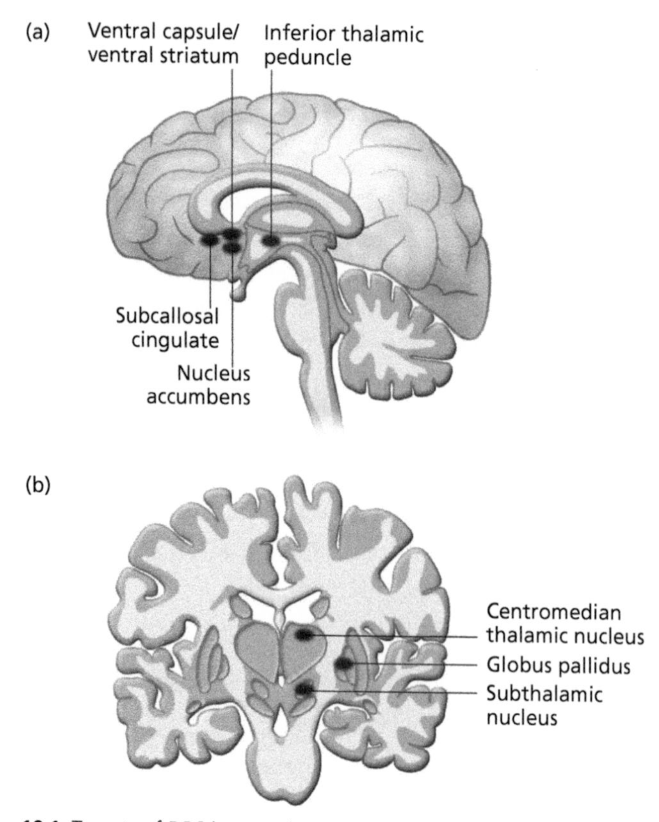

Fig. 10.1 Targets of DBS in neurology and psychiatry.

Reproduced from *The Journal of Clinical Investigation*, 123, 11, Williams NR, Okun MS, 'Deep brain stimulation (DBS) at the interface of neurology and psychiatry', pp. 4546–4556. Copyright (2013) American Society for Clinical Investigation.

This field received a further boost from the field of neuroimaging, but this time from diffusion tensor imaging (DTI) data that allowed for an analysis of the fibre tracts that were affected by DBS protocols as well as the earlier surgical approaches that entailed stereotactic lesions (Schoene-Bake et al. 2010). This analysis showed that the different lesion and stimulation sites, which all had resulted in largely similar proportions of treatment responses, converged onto the medial forebrain bundle (MFB). On this basis a group of German neurosurgeons and psychiatrists decided to target the MFB directly close to its origin in the midbrain. However, because the MFB cannot be detected on conventional clinical (T1 or T2) MRI scans, they had to map this fibre tract in DTI data sets through deterministic tractography (see Chapter 2) before they could insert the electrodes in their pilot study of MFB-DBS in depression (Schlaepfer et al. 2013).

Imaging guided-therapies in psychiatry are still very different from those in areas of medicine where a clear lesion can be detected and targeted. Because it is generally not possible to localize any structural or functional brain abnormalities in individual patients with mental disorders the targets for non-invasive and invasive brain stimulation are largely derived from neuropsychological theories about the underlying networks, which may not do justice to interindividual variability. However, neuromodulation can also act in a different way, by activating or suppressing circuits that are not primarily abnormal but whose modulation may nevertheless produce clinical benefits. Most biological treatments in psychiatry probably follow this path already. For example, monoamine reuptake inhibitors benefit many patients with depression by increasing serotonergic and/or noradrenergic neurotransmission and thus activating compensatory networks, but probably not by correcting an underlying monoaminergic deficit, for which little evidence has been found. An attractive future use of neuroimaging may be in the identification of such compensatory processes and mechanisms of neuroplasticity, which can then be enhanced by the various neuromodulation techniques or by self-regulation training, which will be discussed in Section 10.3.

10.3 How imaging can become a new therapy: the case of neurofeedback

In protocols for self-regulation of brain activity through neurofeedback training, neuroimaging may become a therapeutic technique in its own right. During neurofeedback training, participants receive feedback on their brain activity in real-time and are instructed to change this activation to a desired level. Neurofeedback is an extension of biofeedback techniques, where participants are informed about physiological parameters and try to change them in a particular direction, for example, to down-regulate the heart rate. In the case of neurofeedback, the target signal is recorded from the brain, traditionally with EEG, and patients receive information about a particular EEG parameter, for example, power in the alpha band, in real-time. This information can be provided through symbols on a computer screen, where a change in activation strength can be indicated by the height of a bar or the size of an object. Participants then train to change the brain signal in the desired direction, for example, to increase their alpha power. Because there is often no obvious strategy to achieve this task, they have to try various strategies such as mental imagery or relaxation techniques, and reliable self-regulation, if achieved at all, can require tens of sessions of training.

EEG-neurofeedback has been piloted for a range of mental disorders, but its main current application is for ADHD in children, where a widely used protocol aims at the down-regulation of the ratio between the theta and beta rhythm. The underlying idea is that ADHD is characterized by excessive theta (and/or deficient beta) activity, although this EEG signature has not achieved the status of a robust biomarker. Several clinical trials with neurofeedback in ADHD have demonstrated superior effects compared to attention or concentration training, but whether it is generally superior to control interventions is still a matter for debate and further trials are needed (Arns et al. 2014). Because it has no documented side effects, EEG-neurofeedback is nevertheless a popular technique in the field of ADHD, particularly in the US and several European countries, although its use is less widespread in the UK.

EEG or MEG signals are not the only measures of neural activity that can be used as targets for neurofeedback training. With fast acquisition and real-time analysis of fMRI signals, self-regulation training of neurovascular signals has become possible as well (Weiskopf 2011). In fMRI-neurofeedback (see Fig. 10.2), participants receive continuous information about the activation level in a particular brain area (or network). For some areas, the relationship between neural activity and particular mental activity is well established, for example, the motor cortex in the frontal lobe (motor imagery) or the FFA in the temporal lobe (face imagery). If information about the function of the target area is provided, participants can develop appropriate strategies for successful self-regulation relatively quickly, often even in a

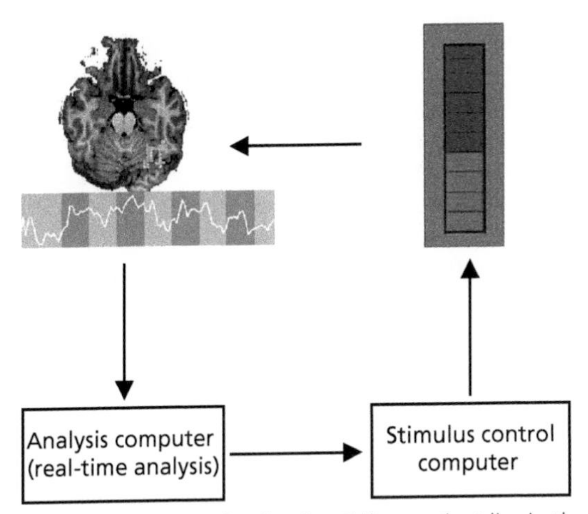

Fig. 10.2 Setup for fMRI-neurofeedback. While a patient lies in the MRI scanner the fMRI signal is processed in real-time, and the extracted statistical parameters (for example, activation strength in a specific brain area) is used to control a stimulation programme, which presents a symbolic (e.g. thermometer) representation of the measured brain activation back to the participant.

single session. Regardless of the higher cost of MRI compared to EEG, fMRI-neurofeedback could therefore become a cost-effective alternative. Furthermore, fMRI can monitor deep brain areas, for example, in the limbic system, more directly than EEG. However, clinical applications of fMRI-neurofeedback in psychiatry have so far been sparse. The author's group has trained patients with depression in the up-regulation of brain areas that were responsive to positive emotional cues over four sessions and reported promising initial clinical effects (Linden et al. 2012). This technique has also been piloted for the treatment of addictions and anxiety disorders, and several clinical trials for the formal evaluation of fMRI-neurofeedback effects in mental and behavioural disorders are underway (Stoeckel et al. 2014).

References

Arns M, Heinrich H, Strehl U (2014) Evaluation of neurofeedback in ADHD: the long and winding road. *Biol Psychol*, **95**:108–15.

Berlim MT, Van den Eynde F, Daskalakis ZJ (2013) Clinically meaningful efficacy and acceptability of low-frequency repetitive transcranial magnetic stimulation (rTMS) for treating primary major depression: a meta-analysis of randomized, double-blind and sham-controlled trials. *Neuropsychopharmacology*, **38**(4):543–51.

Bewernick BH, Kayser S, Sturm V, Schlaepfer TE (2012) Long-term effects of nucleus accumbens deep brain stimulation in treatment-resistant depression: evidence for sustained efficacy. *Neuropsychopharmacology*, **37**(9):1975–85.

Greenberg B, Rauch S, Haber S (2010) Invasive circuitry-based neurotherapeutics: stereotactic ablation and deep brain stimulation for OCD. *Neuropsychopharmacology*, **35**(1):317–36.

Holtzheimer PE, Kelley ME, Gross RE, et al. (2012) Subcallosal cingulate deep brain stimulation for treatment-resistant unipolar and bipolar depression. *Arch Gen Psychiatry*, **69**(2):150–8.

Janicak PG, Nahas Z, Lisanby SH, et al. (2010) Durability of clinical benefit with transcranial magnetic stimulation (TMS) in the treatment of pharmacoresistant major depression: assessment of relapse during a 6-month, multisite, open-label study. *Brain Stimul*, **3**(4):187–99.

Kisely S, Hall K, Siskind D, Frater J, Olson S, Crompton D (2014) Deep brain stimulation for obsessive-compulsive disorder: a systematic review and meta-analysis. *Psychol Med*, **44**(16):3533–42.

Kuo MF, Paulus W, Nitsche MA (2014) Therapeutic effects of non-invasive brain stimulation with direct currents (tDCS) in neuropsychiatric diseases. *Neuroimage*, **85** Pt 3:948–60.

Lefaucheur JP, André-Obadia N, Antal A, et al. (2014) Evidence-based guidelines on the therapeutic use of repetitive transcranial magnetic stimulation (rTMS). *Clin Neurophysiol*, **125**(11): 2150–206.

Linden DE (2014a) *Brain Control.* Basingstoke: Palgrave Macmillan.

Linden DE (2014b) Neurofeedback and networks of depression. *Dialogues Clin Neurosci*, **16**(1):103–12.

Linden DE, Habes I, Johnston SJ, et al. (2012) Real-time self-regulation of emotion networks in patients with depression. *PLoS One*, **7**(6):e38115.

Lozano A, Mayberg H, Giacobbe P, Hamani C, Craddock R, Kennedy S (2008) Subcallosal cingulate gyrus deep brain stimulation for treatment-resistant depression. *Biol Psychiatry*, **64**(6):461–7.

Mallet L, Polosan M, Jaafari N, et al. (2008) Subthalamic nucleus stimulation in severe obsessive-compulsive disorder. *N Engl J Med*, **359**(20):2121–34.

Martinot JL, Hardy P, Feline A, et al. (1990) Left prefrontal glucose hypometabolism in the depressed state: a confirmation. *Am J Psychiatry*, **147**(10):1313–17.

Mayberg H, Lozano A, Voon V, et al. (2005) Deep brain stimulation for treatment-resistant depression. *Neuron*, **45**(5):651–60.

Nitsche MA, Cohen LG, Wassermann EM, et al. (2008) Transcranial direct current stimulation: State of the art 2008. *Brain Stimul*, **1**(3):206–23.

Schlaepfer TE, Bewernick BH, Kayser S, Mädler B, Coenen VA (2013) Rapid Effects of Deep Brain Stimulation for Treatment-Resistant Major Depression. *Biol Psychiatry*, **73**(12):1204–12.

Schoene-Bake JC, Parpaley Y, Weber B, Panksepp J, Hurwitz TA, Coenen VA (2010) Tractographic analysis of historical lesion surgery for depression. *Neuropsychopharmacology*, **35**(13):2553–63.

Speer AM, Kimbrell TA, Wassermann EM, et al. (2000) Opposite effects of high and low frequency rTMS on regional brain activity in depressed patients. *Biol Psychiatry*, **48**(12):1133–41.

Stoeckel LE, Garrison KA, Ghosh S, et al. (2014) Optimizing real time fMRI neurofeedback for therapeutic discovery and development. *Neuroimage Clin*, **5**:245–55.

Weiskopf N (2011) Real-time fMRI and its application to neurofeedback. *Neuroimage*, **62**(2):682–92.

Williams NR, Okun MS (2013) Deep brain stimulation (DBS) at the interface of neurology and psychiatry. *J Clin Invest*, **123**(11):4546–56.

Index

LIBRARY
BMA
BRITISH MEDICAL ASSOCIATION
WITHDRAWN
FROM LIBRARY